Postmodern Sublime

POSTMODERN SUBLIME

Technology and American Writing from Mailer to Cyberpunk

Joseph Tabbi

CORNELL UNIVERSITY PRESS

ITHACA AND LONDON

First published 1995 by Cornell University Press.
First printing, Cornell Paperbacks, 1996.

Printed in the United States of America

∞ The paper in this book meets the minimum requirements of the American National Standard for Information Sciences—Permanence of Paper for Printed Library Materials, ANSI Z39.48-1984.

Library of Congress Cataloging-in-Publication Data

Tabbi, Joseph, 1960–
Postmodern sublime : technology and American writing from Mailer to Cyberpunk / Joseph Tabbi.
p. cm.
Includes bibliographical references and index.
ISBN 0-8014-3074-7 (cloth)
ISBN 0-8014-8383-2 (paper)
1. American literature—20th century—History and criticism.
2. Literature and technology—United States—History—20th century.
3. Postmodernism (Literature)—United States. 4. Sublime, The, in literature. 5. Technology in literature. I. Title.
PS228.T42T33 1995
810.9'384—dc20 94-45473

To the memory of Terry Tabbi

Contents

Preface

The literary criticism of the past few decades, at least in the United States, has inclined toward partial and minimalist forms; it has afforded little room for the un-ironic, expansive gestures that are traditionally associated with the sublime. Equating master narratives, and total conceptual systems of any kind, with political totalitarianism, literary theorists have never been comfortable with writers who speak directly to power. At most, the sublime is allowed to survive today as one critical category among many, as an avant-garde "aesthetic" (Lyotard, "Question" 77) or as a stylistic register in exceptional, "energetic postmodern texts" (Jameson, "Postmodernism" 79), not the least of which are certain texts of postmodern theory.

Despite this critical circumspection, however, the sublime persists as a powerful emotive force in postmodern writing, especially in American works that regard reality as something newly mediated, predominantly, by science and technology. Kant's sublime object, a figure for an infinite greatness and infinite power in nature that cannot be represented, seems to have been replaced in postmodern literature by a technological process. Now, when literature fails to present an object for an idea of absolute power, the failure is associated with technological structures and global corporate systems beyond the comprehension of any one mind or imagination. "Our faulty representations of some immense communicational and computer network," writes Fredric Jameson, "are themselves but a distorted figuration of something deeper, namely the whole world system of present-day multinational capitalism" ("Postmodernism" 79).[1] Jean-François Lyotard, for his part, identifies the pleasure that can derive from the pain of such representational insufficiency as *the* postmodern emotion, a "sentiment of the sublime" that is no less paradoxical than the term *postmodernism* itself ("Question" 77).

The emergence of science and technology has put to flight former metaphysical, religious, and political certainties. As a result, according to Lyotard, we must accept the fragmentation of reality into a set of competing or collaborating "language games." Literature and science in this state of affairs are to be considered alternative constructions or "discourse systems," neither of which should be privileged as a way of knowing. For knowledge to be legitimate, it will have to be local and circumscribed, aware of its own limitations, and suspicious of the global knowledge sought by traditional science. This "incredulity" toward grand narratives will protect us from the illusion that we can "seize reality"; it will "let us wage a war on totality; let us be witnesses to the unpresentable; let us activate the differences and save the honor of the name" ("Question" 82).

For obvious reasons this account of contemporary reality as a linguistic construct has been influential in the humanities, though not especially conducive to their communication with the sciences. I am not working to bring about a common understanding between the two disciplines here, but I do hope to describe the sublime encounter between contemporary literature and science. Downplaying the paradoxes and distancing ironies that both terms of my title seem to invite, I offer *Postmodern Sublime* as a critical investigation of (and occasional polemic on behalf of) a contemporary literary *realism*, one whose psychology expresses itself in the material constructions of an emerging technological reality. The postmodern texts I have selected are challenging in their promise of an affirmative answer to the question that Lyotard and most other constructivist critics never quite get around to asking: With the loss of an absolute standard of reality (assuming such a standard ever existed), and in the absence of metaphysical foundations and former certainties (including the certainty that there is a "nature" that can be represented), have science and technology themselves become means—however "painfully inadequate"—of pre-

1. On the complex sublimity of Jameson's own expression in this passage, the locus classicus of the postmodern sublime, see Redfield (152).

senting the unpresentable, the mind's relation to "the totality of what is" ("Question" 78)?

I concentrate on contemporary novelists—Norman Mailer, Thomas Pynchon, Joseph McElroy, Don DeLillo, and their cyberpunk successors—who invoke the sublime as more than a nostalgic romanticism. Participants in a powerful countercurrent in twentieth-century American writing (running from Henry Adams to Kathy Acker), these writers share an exemplary willingness to push beyond the limits of the literary, to bring their writing into contact with a nonverbal technological reality. The sublime has always located itself between discrete orders of meaning. It is not a category in itself so much as a term that describes what cannot be categorized, and the writers it claims cannot be held to any one literary genre. Thus Adams, in a narrative that is at once a history, an autobiography, and a scientific dissertation, speaks of having been born between two worlds or histories, namely, the idealist eighteenth century of his founding American ancestors and an emerging twentieth-century culture of technological multiplicity. Adams's ambivalence persists in successors as different from one another as Mailer and Donna Haraway, both of whom can be understood (though certainly not reconciled to each other) from a double perspective—that of the romantic sublime and postmodern literary theory.

Like the technological aesthetic it takes as its subject, this book has been of necessity eclectic in its methods, ad hoc and even tendentious in bringing together narrative elements that may well be incompatible. Eric Werner, one of the "Indigo Engineers" interviewed by the novelist William T. Vollmann, speaks of the engineer's need to recycle the products of previous work in constructing a new "machine sculpture," in this case a monstrous hybrid of scrap metal and simulated (or deadly real) organic matter: "Everything I make is a version of something that already exists," says Werner (Vollmann 460). The same can be said of this book, in its use of my own and others' work. Certain sentences and paragraphs may have survived from as far back as my undergraduate days, larger segments from graduate school. Intermediate versions of chapters have appeared in the following journals: portions of Chapter 1 in "Mailer's Psychology of Machines," *PMLA* (March 1991): 238–50; parts of Chapter 3 in "'Strung into the Apollonian Dream': Pynchon's Psychology of Engineers," *Novel* (Winter 1992): 160–80; and portions of Chapter 4 in "The Wind at Zwölfkinder: Technology and Personal Identity in *Gravity's Rainbow*," *Pynchon Notes* (Spring/Fall 1987): 69–90.

I am grateful to these journals (and to the editors of the *Vineland Papers* and the *Michigan Quarterly Review*) for permission to reprint. Excerpts from the manuscript have been read at annual meetings of the Society for Literature and Science, the University of Warwick, the Friedrich-Schiller-Universität (Jena), the 1991 Indiana University conference on interdisciplinarity, the University of Kentucky's twentieth-century literature conference, and Kansas State University's inaugural Cultural Studies Colloquium. The last chapters and final revisions were completed in San Francisco and in Hamburg, Germany, with the aid of a National Endowment for the Humanities summer stipend and a year's Fulbright teaching grant.

I thank William S. Wilson for introducing me to Marjorie Welish's *Small Higher Valley* series of paintings, the first of which is reproduced, with Welish's permission, on the dust jacket of this book. In many ways Wilson's distinctive criticism and Welish's art have informed my understanding of the sublime. At a crucial stage in my revisions, Piotr Siemion generously provided me with a copy of his dissertation, "Whale Songs." Brooks Landon, Larry McCaffery, and Ronald Sukenick gave me the chance to co-edit two "tech-lit" issues of the *American Book Review*, thus encouraging me to write for a nonspecialist audience. For pushing me to think against established disciplinary boundaries—and against myself—I owe debts of personal gratitude and intellectual influence that can be acknowledged only partially within these pages to Tom Adamowski, Linda Brigham, Molly Hite, Tom LeClair, Joseph McElroy, Stuart Peterfreund, and David Porush. Last, I thank my parents for their unconditional support and generosity, and my brother for tutoring me in the latest word processing technology and giving me full use of his home several times during the preparation of the book manuscript.

JOSEPH TABBI

Hamburg, Germany

Introduction

Machine as Metaphor and More Than Metaphor

I am concerned in this book with four of those "most energetic" and ambitious prose writers of recent decades who have sought in their work, in the full knowledge of the impossibility of their task, to reflect imaginatively on the whole of American technological culture (Jameson, "Postmodernism" 79). These writers carry on both the romantic tradition of the sublime and the naturalist ambition of social and scientific realism, but in a postmodern culture that no longer respects romantic oppositions between mind and machine, organic nature and human construction, metaphorical communication and the technological transfer of information. The unprecedented potential for science and technology to assist forms of social, political, and economic control—so often the occasion for an eloquent literary resistance—has at the same time made technology itself a powerful mode of representation. The image of the machine presents faceless and impersonal forces that seem to conflict with the human imagination, but that in their abstraction and precision can also call us outside ourselves. A simultaneous attraction to and repulsion from technology, a complex pleasure derived from the pain of representational insufficiency, has paradoxically produced one of the most powerful modes of modern writing in America—a technological sublime that may be located, conceptually and temporally, between Henry Adams's *Education* and Donna Haraway's "Cyborg Manifesto."

One sentence in particular from the *Education*, in "A Dynamic Theory of History," will serve to introduce the multiple contradictions inherent in any attempt to present the technological culture in its totality:

"The universe that had formed him took shape in his mind as a reflection of his own unity, containing all forces except himself" (475). Adams is here referring to the human being at the earliest stage of self-awareness, to the necessary completion of our primal "education"—the mind's evolution as a recording apparatus—before any historical record can have begun. But the sentence also indicates the imagination's simultaneous separation from and participation in the physical universe, the necessity of its being at once outside and inside its own representations. The sublime emotion before the mystery of the universe is not necessarily that of romantic transcendence, nor is Adams at this moment the romantic figure of the alienated artist individually opposing an objective and impersonal science. Rather, he comes to a self-knowledge that is often explicitly linked to contemporary developments in science, especially the new physics of X rays, statistical thermodynamics, and nuclear radiation. What was true for the writer and artist would be true for the scientist: any representation of the new forces must both contain and exclude the mind of the person representing them; the human subject, inseparable though it may be from the outside world that constitutes it, cannot be integrated into its own symbolic construction of the world.[1]

This is more than an abstract problem in the metaphysics of representation. Adams's self-consciousness reveals the specific situation of an early modernist writer who finds himself separated from the dominant energies of his age, and who must live in an ever more secular culture amid technological forces and emerging corporate systems too complex for any single mind or imagination to know or experience directly. Writing in 1905, the year in which Einstein published his first papers in what were to become the fields of relativity and quantum mechanics, Adams could only anticipate a period marked at every level by discontinuity, when uncertainty would enter into our most fundamental representations of matter and force. Yet Adams had determined, in the face of this uncertainty, to measure all "supersensual" forces, whether physical or historical, by "their attraction on his own mind.

1. W. S. Wilson comments: "The interest of [Adams's] sentence is that the writer or artist has represented the universe as a whole, but the act of representation leaves the artist out of the whole. The problem . . . is that our representations of physical forces don't seem containable within those forces . . . even as the theory of the physicist might not include the working of the brain conceiving that theory" ("Joseph McElroy and Field").

He must treat them as they had been felt; as convertible, reversible, interchangeable attractions on thought" (383).

Not until the mid-1970s, when Joseph McElroy figured the literal growth of a human mind in space, would American fiction get around to exploring the complex self-consciousness within modern science and reflecting that complexity in its form. The appearance of a writer such as McElroy, however, had to wait for the materialization of the scientific worldview, for a moment when technology would make possible transcendence in its digital electronic matrices, in vast computer, transportation, and information networks, and in weightless environments in space and behind the screen of the computer, whose capacity to "translate us toward forms more cerebral" may have had the unforeseen salutary effect, as McElroy remarks in "Holding with *Apollo 17*," of making us *more* aware of our bodies (29). Similarly, when Norman Mailer confronted space technologies directly in *Of a Fire on the Moon*, he claimed to have experienced "a loss of ego" (3), a feeling of dread, disembodiment, and conceptual indeterminacy which he then projected onto the very mechanisms of the *Apollo 11* rocket. Mailer's psychologizing of the machine in *Fire*—a tour de force of metaphorical distortion and transformation—is one way to repair the ego loss that accompanies the loss of "metaphysical, religious, and political certainties" (Lyotard, "Question" 77). Rather than seek some new stability in science or "some virtue of vision" in technology itself (McElroy, "*Apollo 17*" 27), Mailer undermines a science he perceives as monolithic and irresolvably Other; better yet, he assimilates this otherness into his own consciousness, translating technology's mechanical repetitions and otherwise meaningless simulations into a reproductive dream space "where one was his own hero" (*Fire* 160).

Autobiography, and the projection of the self and its immediate sense experience into an autonomous sphere of thought, are continuing responses to scientific indeterminacy and accelerated technological change. Ultimately, however, all such attempts at literary self-creation and technological transcendence must appear quixotic. For just as Adams could not integrate his own mind into the conceptual universe of contemporary science, technology cannot be integrated fully into the mind's symbolic universe. As Haraway would point out in her "Cyborg Manifesto" (1985), the machine is "not man, an author to himself, but only a caricature of that masculinist reproductive dream" ("Manifesto" 152).

The human subject, for its part, has become a "monstrous" nongendered hybrid whose symbolizations exceed all binary categories that separate the human from the mechanical. The subject is now a cyborg, "a kind of disassembled and reassembled, postmodern collective and personal self" that is never whole or natural or innocent, and is not for a moment taken in by "seductions to organic wholeness" such as "bisexuality, pre-oedipal symbiosis, and unalienated labor"—in all senses of the word ("Manifesto" 163, 150).

Typically, the romantic male ego has appropriated the force and terror of woman's labor to itself, as in Adams's famous conversion from the reproductive labor of the virgin to the productive labor of the dynamo. A monstrous birth engendered through technology shows up at the end of *The Armies of the Night*, in Mailer's metaphor of America itself as a woman about to give birth: "and to what?—the most fearsome totalitarianism the world has ever known?" (316) In every case the detached male ego tries to achieve embodiment through a maternal Other, whose reproductive force is conceived in opposition to the technological absolute.[2] (Doubly disturbing for Mailer, then, is the prospect of women joining *with* technology in their own efforts at liberation, a prospect that haunts him throughout *The Prisoner of Sex*.)

Haraway's radicalism lies precisely in her rejection of this "masculinist reproductive dream" (although she does not go so far as Patricia Yeager, who imagines the labor of giving birth as itself the foundation of a "maternal sublime" capable of displacing transcendent, disembodied versions of the masculine sublime). Embodiment for Haraway is technological; the machine is the replicative force behind social institutions and is thus a powerful ally against transcendental romanticism in all of its forms. But hers is necessarily a *partial* alliance, limited to local interventions in the present technological order and forever "ironical" in its efforts at self-creation. The cyborg would resist all totalizing dynamics as well as the more futile strategies of literary opposition; it would allow no system to dominate everything, no dialectic or "final appropriation of all the powers of the parts into a higher unity" ("Manifesto" 150). But Haraway's resulting commitment to a world constructed of localized "differences" does not prevent the continued construction of total systems and the transcendental subject; and the assertion of difference has scarcely affected the prestige that

2. Yeager understands Mailer's metaphor of the woman in labor as representing "a force that is unstoppable and awesome, even as she yields in the final paragraph to the exhortations of male power" (18). Mailer calls out to the Vietnam war protesters to "'rush to the locks' [*Armies* 317], to surmount the terror of the unruly body-politic that his laboring woman represents" (17).

a mostly monolithic science continues to enjoy. The cyborg is a product of global technology after all, and any local difference it makes can scarcely be expected to alter this global context.

Haraway herself admits as a " 'final' irony" (the scare quotes are hers) the possibility that "the cyborg is also the awful apocalyptic *telos* of the 'West's' escalating dominations of abstract individuation, an ultimate self untied at last from all dependency, a man in space" (150–51). And at times it is difficult to separate the most oppositional of romantic self-projections from the transcendental dreams of technology itself. Consider, for example, a statement by the German rocket engineer Wernher von Braun, cited by Thomas Pynchon as an epigraph to the first section of *Gravity's Rainbow:* "Nature does not know extinction; all it knows is transformation. Everything science has taught me, and continues to teach me, strengthens my belief in the continuity of our spiritual existence after death" (1). Mailer, too, had come across a variation of this statement—he circled it and underlined it twice—in a NASA press release during the week of the *Apollo 11* launch.[3] Both novelists obviously share von Braun's concern with ideas of continuity, Mailer in his ceaseless imaginings of "death as a continuation, a migration, a metamorphosis" (*Cannibals and Christians* 325), and Pynchon in his attempt to imagine all aspects of organized human activity, be they scientific, political, economic, aesthetic, or sociological, as capable of being transmuted somehow to a psychic world "beyond the Zero" (*Gravity's Rainbow* 1). However ironically such deathless projections are to be taken, they nonetheless reveal the writer's necessary complicity in the construction, and a new legitimation, of the technological culture.

There are of course obvious difficulties with von Braun's dream of "continuity," for it depends on an uncritical transference of a technological worldview (which is questionable in itself) to a wholly separate

3. NASA press releases and flight transcripts may be found in abundance among Mailer's papers, which he keeps stored in a warehouse on East Sixty-second Street in New York. I was allowed access to these papers through the courtesy of Robert Lucid, Mailer's archivist and biographer. The von Braun quotation is from Simon and Schuster's *Third Book of Words to Live By* (1962), a volume that was widely quoted by NASA during the month of the first moon launch (July 1969). On the opening page of the von Braun press release Mailer has underlined the sentence "Immortality is a continuity of our spiritual existence after death." The circled passage appears on page 3. Those sentences that Mailer underlined read: "Nothing in nature, not even the tiniest particle, can disappear without a trace. Nature does not know extinction, all it knows is transformation. If God applies this fundamental principle to the most minute and insignificant parts of his universe, it just does not make sense to assume that he would not apply it to the masterpiece of his creation, the human soul."

realm of experience—the development of a conscious human life. The engineer treats life not as a form but as an accumulated sum, where "nothing in nature, not even the tiniest particle, can disappear without a trace" (von Braun press release 3). But what of the organizational process that gives the particles direction and significance, whether one chooses to conceive of such activity as a traditional teleology, the work of an all-knowing and prescient god, or the self-organizing activity that is thought to characterize living systems in nature? Related to this more or less formal difficulty is a rhetoric that can scarcely stand up to the most basic political scrutiny. For, contrary to the comforting statement given out for publication at the launch of *Apollo 11*, the technological future that von Braun himself had worked to realize could hardly have grown out of an objective lesson taught by science, any more than his past career as the leading engineer at the Peenemünde rocket facility would have been possible had he continued alone in a youthful apolitical quest for knowledge.[4]

I mention these historical and biographical details not to pass judgment on von Braun's—or Walter Dornberger's or (in light of recently declassified British Intelligence documents) Werner Heisenberg's complicity in, and later silence on the subject of, Nazi crimes.[5] Any denunciation of a fifty-year-old evil, coming from an American born two full generations after von Braun, would be facile at best; at worst it might obscure the quite different technological and market relations that have come to be perceived by many as comprehensive and all-powerful today. The form of postmodern domination (what Slavoj Žižek might term the "sublime object of ideology") tends to manifest

4. After the Second World War, engineers such as von Braun and his superior, General Walter Dornberger, were clearly anxious to downplay their role in German political history. For example, only once in his 1954 memoir does Dornberger mention the "foreign laborers" at Peenemünde, not as the German Jews and Russian prisoners of war who were forced, under his command and in deadly conditions, to labor on the V-2 rocket, but as the apparently expendable group whose barracks suffered the most damage in the Allied bombing of the rocket works. Molly Hite notes Dornberger's omission in *Ideas of Order* (148), and Steven Weisenburger documents Pynchon's Peenemünde source material in *A "Gravity's Rainbow" Companion*.

5. Heisenberg and nine other nuclear scientists were held for questioning after the Allied troops occupied Germany, and their private conversations were recorded by British eavesdroppers and eventually published as the "Farm Hall Transcripts" in 1992. In a review of these transcripts David Hollinger writes, "Although the transcripts contain a handful of direct references to Nazi atrocities, Nazism figures in these pages more as an inconvenience than a crime," so caught up were these men in scientific professionalism and nationalism. Hollinger excepts only "the legendary anti-Nazi von Laue," but not Heisenberg (32).

itself not in offensive individual attitudes that are easy to denounce, but in large forces of corporate organization that control the social and economic relations among human beings. Such domination should not be imagined as producing a "false consciousness" in its subjects or distorted representation that hides the truth behind misleading appearances; nor should we suspect a technocratic elite or overt censor of attempting to impose compliance "from above" on a deluded majority. As Noam Chomsky has pointed out, the majority nowadays are rather more likely to be influenced by an "elite consensus" of right-thinking people more or less like anyone else, but reinforced in their rhetorical stance by sophisticated (and, for most people, hardly perceptible) techniques of media persuasion, statistical analysis, and "public opinion" control.[6]

The most effective and potentially dangerous ideological force *at this moment* is to be found in those things we do every day to sustain the technological culture, whether what we do is explicitly technological or not, unconscious and internalized or, if conscious, sweetened with a suitably hip postmodern irony. For it is true that most individuals—at least those with any real economic power or professional calling to political articulateness—have become far too sophisticated to be caught out in earlier forms of false consciousness. This sophistication and a new self-censorship may explain why the same self-conscious irony that helped make for such all-embracing social and political criticisms in *Gravity's Rainbow* has by now frustrated any direct political opposition, as Pynchon's 1990 novel *Vineland* all too clearly shows. The internalization of dominant systems of oppression had its classic American expression, independently of Michel Foucault's prison book (and four years before that book's appearance in English), in this universalizing sentence from *Gravity's Rainbow:* "The Man has a branch office in each of our brains, his corporate emblem is a white albatross, each local

6. In *Manufacturing Consent* Chomsky writes: "Most biased choices in the media arise from the preselection of right-thinking people, internalized preconceptions, and the adaptation of personnel to the constraints of ownership, organization, market, and political power. Censorship is largely self-censorship, by reporters and commentators who adjust to the realities of source and media organizational requirements, and by people at higher levels who are chosen to implement, and have usually internalized, the constraints imposed by proprietary and other market and governmental centers of power" (300). Christopher Norris, in *Uncritical Theory*, discusses the (superficial) similarities of this conception to Michel Foucault's postmodern theories of the internalization of contemporary forces of "power/knowledge" (100–121). The crucial difference between Foucault and Chomsky is that the latter does *not* perceive this process as total and ineluctable.

rep has a cover known as the Ego, and their mission in this world is Bad Shit" (712–13). By the time of *Vineland* this essentially psychological model of power had been *externalized* and made automatic by advances in information technologies. As one character points out (in a cynical rationalization of his own compromises and betrayals of his former revolutionary ideals): "Everybody's a squealer. We're in th' Info Revolution here. Anytime you use a credit card you're tellin' the Man more than you meant to. Don't matter if it's big or small, he can use it all" (74).

In both quotations Pynchon is concerned with the absorption of alternative and oppositional tendencies by established systems of domination; both present power, and "our" complicity in it, as something total.[7] One strength of *Gravity's Rainbow* was its use of irony to cut through the hypocrisy and false consciousness of a conformist generation that had (in the eyes of many sixties radicals) repressed most personal knowledge of the way power gets internalized. But in *Vineland* irony is now used, by this very hip character and his wife, Frenesi Gates, to *support* the dominant system, to legitimate a world in which complicity is assured through technology, not psychology. If "everybody's a squealer," and everybody *knows* it, then irony itself becomes the dominant cultural attitude, part of the political problem, not the solution, and a hindrance—on the aesthetic level—to any attempt to ground fiction in some shared "human reality" and in a psychological authenticity, which Pynchon feels to be missing in some of his undergraduate fiction, "found and taken up, always at a cost, from deeper, more shared levels of the life we all really live" (*Slow Learner* 18, 21).

To a novelist of the generation after Pynchon's such as David Fos-

7. This is the totalizing viewpoint, of course, of only one character in *Vineland*, and not an especially sympathetic character at that. In the novel as a whole, as Molly Hite has shown, "complicity is not by definition total and does not by definition rule out resistance" ("Feminist Theory" 148). Hite is responding in part to my own review of *Vineland*, in which I charged Pynchon (in terms that were surely too sweeping) with failing to resist the commercialized American culture of simulation that is his subject in the novel ("Pynchon's Groundward Art"). Hite offers a convincing account of an alternative "political sensibility" in *Vineland*, "one that tacitly acknowledges the possibility of bad faith in subsuming considerations of exploitation to considerations of mortality" ("Feminist Theory" 148). There can be no resistance when "the Man" sits in each of us as absolutely as death. Thus, what I initially understood as a failed resistance might be better seen as an attempt by Pynchon to imagine a strategy of *disinternalization*, one that remains critical but refuses to meet power on its own ground, even to oppose it.

ter Wallace, self-consciousness and postmodern irony, though once "useful for debunking illusions," have by now become "poisonous," at best "a measure of hip sophistication and literary savvy. Few artists dare to try to talk about ways of working toward redeeming what's wrong, because they'll look sentimental and naive to all the weary ironists. Irony's gone from liberating to enslaving" (McCaffery, interview 146, 147). The media culture has absorbed its own strongest critiques, frustrating any direct resistance grounded in traditionalist conceptions of realism and existential authenticity. Pynchon's stated desire for such realism, which has the merit of risking the appearance of naïveté that Wallace feels is needed now, does not free Pynchon from the grip of irony. To function as a mode of social and political critique, irony requires just the foundation Pynchon speaks of in a shared social and psychological reality, a ground from which it can operate and a set of certainties it can then go on to annihilate. In a postmodern period that questions certainties as a matter of course and, in the absence of totalities, fixed values, and absolute conceptual foundations of any sort, irony loses its oppositional force, or is reduced to cynicism and black humor. To adduce further ironies in the context of this postmodern condition is not subversive but redundant.

Žižek speaks for an un-ironic postmodern realism when he argues that "ideological propositions" do not go away simply because people cease to take them seriously (33). His Lacanian analysis of corporate capitalism, in which fantasy is an enabling presence *within* real processes of corporate control and not in the least opposed to them, has much in common with my analysis of an "engineering psychology" undertaken later in this book. There von Braun's dream of immortality will be seen as a fantasy projected onto the very objects and methods of engineering technology, which are in turn imagined to be animate, independent of human control, and capable of an autonomous development after those who created the technology have left the scene. Such a transcendental technology has its economic correlative in the abstract processes of commodity exchange discussed by Žižek, and in a culture where technological advances come to represent advances in consciousness even though such technological processes need not be grasped, and often could not ever *be* grasped, by any one human mind. In both cases individual beliefs and desires externalize themselves in technological inscriptions, and consciousness is delegated to symbolical systems—the "hard" memory of archives and computer networks, for example, or to processes of economic "growth" and commodity exchange which are not thought, but which in their essential abstraction have the *form* of thought.[8]

In this respect McElroy's conceit in *Women and Men* of an international economy that converts thought electronically into credit is something more than one intellectual's self-validating science fiction fantasy. Like William Gibson's consensually hallucinated cyberspace, the conceit is simply a literalization of the enabling fantasy behind money. That is, we know that money, "in its material reality," is a wholly symbolic expression of our social relations, but in practice we act *as if* it were "the immediate embodiment of wealth as such," and so we assign a material reality to a sphere of action that is "in reality" wholly simulated (Žižek 31). Žižek speaks of this "sublime object" as neither wholly mind nor wholly body but an "immaterial corporeality" (18). McElroy in an important early essay speaks similarly of his own narratives as "concrete abstracts" of the consciousness that is fixed in buildings, transportation systems, and the entire mass of life in a contemporary city ("Neural Neighborhoods" 201). In both cases reality is irreducibly decentered and externalized; it is located not in any one human mind or body but in the social relations conducted *among* human beings through various simulations and abstractions, through bureaucratic institutions, and through the machines that enact "the automated thoughtfulness of the community" (*Women and Men* 235). To represent reality must be at least in part to show what this abstraction *feels* like, and readers who mistake felt abstraction ("conceptual to the touch") for the private expression of the author's mental life are unlikely to get much pleasure from McElroy's novels (*Women and Men* 296).

Žižek, among many other critics of postmodernism, has also noticed that the mechanisms of corporate capitalism have by now evolved to the point where they are themselves a kind of avant-garde, fomenting a state of constant revolution in international markets that is not to be "resisted" by the subtlest postmodern irony or the most inventive of aesthetic intelligentsias. Like the postwar multinational economy that it facilitates ("read 'bottom-line American,'" says one of the polyvocal angels in McElroy's *Women and Men*), global technology depends on incessant development "to resolve again and again, come to terms with, its own fundamental constitutive imbalance, 'contradiction'" (Žižek 52). In particular, Žižek argues, discrepancies between

8. This is Žižek citing Frankfurt School critic Alfred Sohn-Rethel (19).

the "social mode of production and the individual, private mode of appropriation," a collectivist market and an advertising cult of individuality, produce aggravations in the capitalist system that, far from disrupting it and ultimately causing its collapse (as Marx had predicted they would), have proven necessary to its continuation (52). To be sure, this recognition of capitalism's staying power and its ability to neutralize its own most powerful critiques did not have to wait on Žižek's postmodernist, post-Stalinist critique. Scarcely a generation after the death of Marx, Adams could already wonder at the capacity of a society that "cared more for chaos than continuity" to avoid the suicide course on which it appeared to be headed, and to save itself from anarchy "by its vitality, by its prodigious capacity for successive ruination, or by the discovery of a new and available source of power" (Blackmur 8).

The exceptional dynamism of capitalism could be said, and has been said by many, to have produced a corporate culture that is itself sublime, an economy whose contradictions are all too conducive to those contradictory feelings—"pleasure and pain, joy and anxiety, exaltation and depression"—that have been traditionally associated with the sublime: "There is something of the sublime in capitalist economy," writes Lyotard, who with Žižek and Jameson regards this sublime economy as a logical outcome, even an apotheosis, of more than two centuries of philosophical idealism ("The Sublime and the Avant-Garde" 246). Capitalism, in Lyotard's view, "is, in a sense, an economy regulated by an Idea—infinite wealth or power. It does not manage to present any example from reality to verify this Idea. In making science subordinate to itself through technologies, especially those of language, it only succeeds, on the contrary, in making reality increasingly ungraspable, subject to doubt, unsteady" (255).

The constantly deferred utopia of a market economy leaves consumers with an unreal sense of history, and this is not a situation from which the most "historical" of literary imaginations—even Adams's or Pynchon's—can easily separate itself.[9] The demands of technology (and the markets it creates) for continued innovation, for "modernization" and constant "improvement," cannot be separated from the modern writer's need to find a new style for each project, to be "original," and, in the absence of any stable value or widely shared belief system, to create ever novel mechanisms of literary form. The metaphysical

9. "The complicity of modernism with consumer capitalism," writes John McGowan, "is a favorite postmodern theme, found in writers as diverse as Graff

loss of continuity finds its literary counterpart in a heterogeneity of style and genre, and in the aesthetic necessity not only to differentiate oneself from other writers and artists but to recycle the themes of one's own earlier work in the light of constantly changing historical conditions. Ambition and the desire to surpass oneself and one's aesthetic precursors have long been characteristic features of the sublime, whether the relation is conceived in Oedipal terms or in terms of the literary history of supplanted aesthetic techniques. (Even Longinus, as long ago as the first century A.D., could identify numerous defects in attempts at sublime expression as arising from "the same cause, namely, a desire for novel conceits, the chief mania of our time.") But literary ambition and the cult of originality has reached hitherto unheard-of intensities in a technological age where performance, the conversion of mind and raw talent into a visible and negotiable product, is the only recognized value.

The sheer virtuosity of the four novelists discussed in this book has inspired their best commentators to use terms of aggrandizement in describing their work—as "encyclopedic narratives" (Edward Mendelson) or as a "literature of excess" and "mastery" (Tom LeClair) that would match the sheer productive power of the larger culture. These critiques have been instructive, informed as they are by many of the "nonliterary" discourses that sustain the technological culture in the first place. But the appearance that such discussions engage directly with the technological subject, that they and the novelists under discussion have somehow "mastered" the material of the technological culture, is ultimately illusory.[10] It is not enough to say, as LeClair does,

(1979), Newman (1985), and Foster (1985)" (11n). My own discussion of this theme draws on these writers as well as Sande Cohen's discussions of consumer culture in *Academia* and Richard Poirier's essay of 1987, "Venerable Complications: Literature, Technology, and People." See chapter 7.

10. Like Mailer (the writer least in tune with the "systems novelists" he studies), LeClair is still committed to an oppositional literature whose "excess represents and critiques excess" (17) in the dominant culture when the real challenge, I think, is to walk away from domination, to imagine a fiction—and so create a sensibility—that is free of the conventional dialectics of power and its critique. Haraway may be no closer than LeClair to naming a nondialectical politics of (in her case feminist) resistance; she does not, any more than LeClair, "go beyond professions of hope and the critique of domination" (Csicsery-Ronay, "SF of Theory" 397). Her field network, like LeClair's systems theory and William S. Wilson's postmodernist "field," remains largely caught up in the present ideology it seeks to supplant. But Haraway's conception of freedom within "a transformed field [that] sets up resonances among all its regions and components," however hypothetical, employs a

that excess and complexity are necessitated by the complex world we inhabit, unless one assumes that an order already exists in the world which can be captured through a mimetic art, one whose "fundamental . . . accomplishment [is] the creation of imitative forms" (*Art of Excess* 23). The excess to be found in technological systems (such as Mendelson's encyclopedic technology of reference) is scarcely of a type with literary excess; the individual writer, often a loner with pen and paper, could never compete with the high-budget productions of the various corporate media. Excess in this fiction is not simply *more* than but *other* than the technological mechanisms, media, and categories it deforms. It occurs to the young critic in DeLillo's *Mao II* (a character who at one point in the novel happens to be associated with LeClair) that the work that "we have in front of us represents one thing. How we analyze and describe and codify it is something else completely" (222). A semiotic marker, excess can be said to represent a surplus of signification that always separates the real from its symbolization, the work of art from the master narratives it occasions.

What the sublime "exceeds," in the end, is the very possibility of symbolization, although this overabundance is hardly a reason to do away with symbolic categories, knowledge systems, and sets of assertions about what we can reasonably put forward as true. Narratives that are already in place in the world are not about to disappear because they fail to accommodate excess, nor are they likely to be changed solely through cognitive skepticism or the purportedly subversive force of irony—attitudes that, in their more thoroughgoing textualist forms, seem to me to be as aesthetically limiting as they are politically disabling. Pynchon in particular (the writer treated at length in this book who is most often claimed by the postmodern textualist camp) has suffered on both counts, the political and the aesthetic, from readings that routinely regard the absence of determinate meaning in his novels as their only significance.

Of course, the powerfully significant failure to signify has always characterized the rhetoric of the sublime, and there is much in Pynchon's novels to support ironical, indeterminist readings. Yet the intersecting worlds, absent centers, and dissolving categories in his work, destructive though they may be of the status quo (and of the reader's peace of mind), are not wholly negative in their implications. Rather,

---

noncontestatory language that remains oriented toward a future, not set against an admittedly outdated paradigm (Csicsery-Ronay 394, citing *Primate Visions* 5).

when they are successful, such figures disturb our complacency. They encourage us to look for meaning *between* systems and at the point where categories break down, and not passively to accept indeterminacy, ambivalence, and an only superficially subversive irony at face value, or as values in themselves.[11]

From an aesthetic standpoint the case against irony is put succinctly by Harold Bloom when he notes that "ambivalence, increased to excess, becomes irony, which destroys the sublime" ("Foreword" ix). This sentence appears in Bloom's foreword to *The Romantic Sublime* by Thomas Weiskel, another foundational book whose fine-grained distinctions have made it possible here more or less accurately to identify occurrences of the sublime moment in four contemporary writers, if not to associate them with a prestigious tradition, a received "canon," in romantic literary history. One cannot have a canon when the most valued literary energies dissolve fixed orders and destabilize systems, including those aesthetic orders that, in LeClair's words, promote "a kind of literary transcendence" in their founding criteria of "permanence and universality" (*Art of Excess* 17). That such criteria have lost their appeal even for the contemporary writer who stands to benefit the most from them will be seen in the section of chapter 7 that follows the fascinating spectacle of DeLillo, in *Mao II*, actively resisting his own attempted inclusion, by LeClair and no less influential a critic than Frank Lentricchia, in the postmodern pantheon.

Admittedly, with Lentricchia and LeClair, I am making a case for reading texts by DeLillo more often and more seriously, and my argument is based partly on the recognition of a powerful technological aesthetic that unites DeLillo (and McElroy and Haraway and Kathy

11. In *Dissident Postmodernists* Paul Maltby notes the tendency of many Pynchon critics, when they comment on his literary use of informational entropy, "to disregard the *adversarial* value of this practice and discuss only its aesthetic or philosophical significance in apolitical terms. . . . Pynchon's objective is to fashion a discourse which maximizes the effects of ambiguity, indeterminacy, and paradox *with a view to occupying the excluded middle*" of positivist logic (145–46). By using excessive language, Pynchon establishes his own meaning systems in the conceptual spaces between logical alternatives, at the edges of positivist orders of knowledge, and in the inexplicable passages between orders of discourse. Maltby later notes the "strategic importance" that Pynchon attaches "to a whole domain of *nonverbal* signs, specifically those marginal ones that have been 'passed over' by the System" (171). This "*ideal* of the revelationary nonverbal sign" is very much in keeping with Pynchon's negative, critical version of the sublime and its continued resistance to rational systems building of *any* kind (171). Maltby remarks on the limits of this wholly oppositional mode when he discusses Pynchon's counterforce, who are "not only effortlessly absorbed by the System, they are understood to sustain it" (161).

Acker) with more established writers such as Mailer and Pynchon, as well as with other risk-taking and dissenting postmodernists such as Robert Coover, William Gaddis, and (in England) Nicholas Mosley and Christine Brooke-Rose. My particular grouping of writers is not, I hope, a "closed" system that would exclude other writers, nor am I interested in making their undeniable technical sophistication a value in itself. Rather, these writers are of interest to me insofar as they bring the full range of postmodern techniques in fiction to bear on the larger culture. All are in fact "high" postmodernists, but all (with the unfortunate exception of McElroy, who deserves to be better known) have found an audience in the corporate culture within and against which they write. Most important, given my master trope in this book, these are the writers whose "energetic postmodern texts," in Fredric Jameson's words, "afford us some glimpse into a post-modern or technological sublime" ("Postmodernism" 79).

Jameson, however, has had little time to spare for this fiction. He has gone on (and critics such as Larry McCaffery, Scott Bukatman, Brooks Landon, Veronica Hollinger, and Marleen Barr have followed him) to isolate cyberpunk fiction as "the supreme *literary* expression if not of postmodernism, then of late capitalism itself" (*Postmodernism* 419n). I prefer to concentrate on the group that Jameson neglects, in part because their fiction is more to my taste, but also because it is not yet as dependent as cyberpunk (and the earlier radical feminist science fiction celebrated by Haraway) on nonliterary corporate media for their "literary expression." Most science fiction, as Bukatman admits, continues to be "unflaggingly conservative in its language and iconography," or at best reflective of its era's "language of spectacle and stimulation" (6, 11). No discourse is autonomous, but when fiction can reach the sublime register only by borrowing symbolic capital, copyrighted effects, and communicative energies from other media, it is likely to be the losing partner in a corporate merger, as I argue in the epilogue.

Thomas Weiskel, writing in the early seventies, does not actively question the canonical status of his chosen authors, though he does mention in passing that the sublime moment would seem to have "brought the high and the low into dangerous proximity" (19).[12] Weis-

12. Weiskel, however, would revise the canon slightly to include William Collins, who is less often read today than his romantic successors Wordsworth, Blake, Keats, and Shelley. A writer whose purported obscurity "resides in 'the uncommonness of the thought,'" Collins is the McElroy of the romantic sublime (Weiskel quoting Samuel Johnson, 110).

kel's mode of reading, his aesthetic distinctions, and even his moral evaluations are far from normative, and he demonstrates an admirable ability to take his categories seriously without allowing them to delimit thought or fix the terms of later inclusions in the canon. The egotistical sublime, the negative or Kantian sublime, the mathematical, the metaphorical, the metonymic, and the dynamical sublime—all may be seen to function, to varying degrees and at different moments, in the work of Mailer, Pynchon, McElroy, and DeLillo.[13] But apart from the precision and undeniable descriptive power of the categories, these authors prove to be most sublime—as Weiskel himself again and again approaches a critical sublimity—at precisely the point where the proliferating categories break down, giving way to an all but irrepressible heteroglossia (the result, in Bakhtin, of any literary form that continually and honestly tests itself against a "developing reality" [*Imagination* 39]). We might even go so far as to view Weiskel's categories as themselves symptoms of the overabundance of signification, in the world and on the page, which romanticism discovers and the technological culture intensifies. If this is so, we should look not to any one category or system to describe the scene of contemporary literary power but (again) to the edges of systems and the spaces between categories, where one discourse or sign system interacts with other, potentially homologous but necessarily separate systems.

One could hardly find a better contemporary occasion for the sublime than the excessive production of technology itself. Its crisscrossing networks of computers, transportation systems, and communications media, successors to the omnipotent "nature" of nineteenth-century romanticism, have come to represent a magnitude that at once attracts and repels the imagination. Indeed, the pattern of attempted apprehension and imaginative collapse that Weiskel details in nineteenth-century engagements with nature can be seen in our own imaginative engagements with technology. "The sublime appearance promises an overabundance of stimulation," to which the imagination can react passively or actively: either the imagination wishes to be inundated in the

13. I have purposely omitted one of Weiskel's categories, the theological sublime, since the postmodern sublime of Mailer, Pynchon, McElroy, and DeLillo represents the fate of narrative in a wholly secular time, in which any numinous object or publicly sharable system of uniform belief has been replaced by a belief in a polymorphous "Other," the collectively produced object world as it is materialized through technology.

network, and thus risks experiencing a loss of identity or "anxiety of incorporation," or it desires to oppose or replace the sublime appearance with a linguistic construction of its own, to "possess" verbally the object of its anxieties (Weiskel 105).

The verbal construction, however, is never adequate to the technological object; the mind that would conceive itself as energy and force is dwarfed in relation to technological force. The resisting imagination is then left with an excess of meaning, an "excess on the plane of the signified" (to use Weiskel's phrase out of structural linguistics) which cannot be simply added on to a verbal system that is already known (137). Instead, the technological network, like the natural world and Adams's universe of force, remains separate, unfamiliar, *other* than the semiotic system. (In a Pynchonesque intersection of mathematical and verbal systems, a semiotic shadow, a "delta t," always stands between words and what they stand for.) It is in this realm of excess signification at the boundaries of discourse, the point at which meaning is overwhelmed and entropy threatens to take over, that the technological sublime, to append yet another term to Weiskel's list, begins to function in literature.

The form that so paradoxical a signification will take cannot be given in advance as a one-to-one mimetic matching of technological mass and semiotic excess. As Weiskel notes in a critique of the eighteenth-century Lockean psychology on which much of the nineteenth-century romantic sublime was based, it is not true that "*vast* in one domain or discourse simply equals *vast* in another" (15). Technological accumulations are not to be grasped by the most formally inventive or encyclopedic narrative; still less are they to be simply *opposed* by the force of imagination, whether the writer engages in outright war on technology or cleverly attempts—as Mailer often does—to subsume its power as one subordinate term in a creative dialectic. In Mailer's highly idiosyncratic style of existentialist poetics, technology is to be rescued from its Sartrean status as the dead repository of alienated labor. Through the willfully irrational psychology of machines (reminiscent at times of Sartre's psychoanalysis of things), the object world of corporate NASA technology is triumphantly resisted, having been perceived by the author as "totalitarian" and potentially dominating in the first place. But such resistance is only a stage in the unfolding of a larger project in which the technological world is to be made a part of the subjective mind. The imagination deals with excess, but by recognizing it as "other" and then going on to include it within its own purportedly larger perspective. In this way the conservative romantic imagination is kept intact; it attenuates its own fear of identity loss or incorporation by the technological order, but at the cost of introducing "dread," anxiety, and ontological insecurity into the very materials and mechanisms of technology itself.

Technology in Mailer is thus transmuted into Self, producing a more aggressive, existentialist version of the "egotistical sublime" that Keats objected to in the early poetry of Wordsworth and that Weiskel generalizes to describe any poetics that seeks—whether consciously or not—to replace the incomprehensible, sublime object with something else (namely, an image of the perceiving mind and imagination). It might be said that, with *Of a Fire on the Moon*, *Plus*, and *Gravity's Rainbow*, twentieth-century American literature has come to fulfill Wordsworth's famous prognosis in the "Preface" to *The Lyrical Ballads* that the eventual function of the poet would be to "carry sensibility" into the very heart of scientific reality. Yet the postmodern technological culture has proven all too efficient at turning aesthetic sensibilities to use and profit. The sublime desire to assimilate external reality into the substance of mind is not so different after all from more collective modes of aggrandizement, where wealth and knowledge accumulate as capital, and—in its later, postmodern phases—capital begins to take on the form of a dream by attaching all desire to itself.[14] To see where Wordsworth's aesthetic was heading, one need only look to the production of popular science fiction in the decade of Jameson's essay "Postmodernism, or the Cultural Logic of Late Capitalism," especially the "new romanticism" of William Gibson's cyberspace trilogy. The lyric self that once sought absorption into the attributes of God has by now been incorporated into Capital. And what began as a hidden

14. Weiskel cites as an example Wordsworth's "I Wandered Lonely as a Cloud," in which "everything external or 'out there' is transmuted into the substance of mind, which accumulates like a kind of capital. 'I gazed—and gazed—but little thought / What wealth the show to me had brought' " (52–53). Thomas Schaub has written on the "language of capital acquisition" in Mailer's foundational essay "The White Negro" (*Cold War* 158), and Jameson anticipates Schaub in an early essay on *Why Are We in Vietnam?* (The classic account of capital's symbolic occupation of the human psyche is of course Jameson's later book *The Political Unconscious*.) Far from resisting the "consensus" culture of liberal capitalist society (as Mailer would have it), *Vietnam* represents, according to Jameson, "some hypostasis of *competition* itself as a social and historical mode of being" ("The Great American Hunter" 193). See chapter 2.

strategy of romantic opposition (becoming ever more ironical as the imagination's pretensions to mastery, disembodiment, and autonomy become more and more clear) has ended, in Pynchon's and Mailer's cyberpunk successors, in a corporate merger.

For the postmodern ego which goes on romantically asserting its independence from all technological determinations, the price continues to be alienation and a deep self-division. Resistance in Mailer, for example, ends up being directed not just at the so-called totalitarian constructions of the technological society. For, as we shall see, his work is marked by a ceaseless supersession of *his own* earlier accomplishments, and the constructed self is forever divided against itself in the ongoing attempt to exceed the limitations of its earlier versions. The form taken by Mailer's life-narrative over the course of a long career has been repetitive and circular, or at best an Emersonian spiral that continually ascends by the energy of its own oppositional extremes. In its desire to subsume all these oppositions and self-contradictions into a single movement of thought, it is unapologetically dialectical. But the difficulty with all such dialectical resolutions is that they tend to aggrandize self-consciousness at the expense of otherness, be it social, natural, or the objective otherness of the technological, collectively constructed life-world. For this external, incommensurable vastness the mind substitutes its own linguistic infinity and so identifies two categorically separate realms in a willful act of the imagination, a resolution that is at best metaphorical.

From the contemporary perspective of Haraway's cyborg metaphysics, a feminist "theory of 'difference' whose geometries, paradigms, and logics break out of binaries, dialectics, and nature/culture models of any kind" ("Marxist Dictionary" 129), Mailer's conceptual oppositions and the persistence of an antitechnological rhetoric lose much of their force: "The dichotomies between mind and body, animal and human, organism and machine, public and private, nature and culture, men and women, primitive and civilized are all in question ideologically" ("Manifesto" 163). Even metaphor, that ultimate literary weapon against the depressing world of facts, has been called into question as a way of thinking about technology, since a technological construction may itself already be an embodiment of human thought processes (even as the universe of force, in Adams, already contains the mind that represents this force). The very word *metaphor*, originally "to carry across," implies separation and a distance to be traversed, and it has always been a powerful way of bridging (or perhaps merely papering over) the break between two separate orders of discourse. But if one no longer maintains the old romantic dualities, there will no longer be this need either to oppose or to unify technology and the imagination. The writer may instead regard technology in its own element, not as a resistant object world that needs to be "humanized" through metaphor, myth, or a personal literary style, and not as "an *it* to be animated, worshipped, and dominated," but as some larger, unattainable version of "us, our processes, an aspect of our embodiment" ("Manifesto" 180). In such a narrative the machine itself emerges as a metaphor, a figure representing forces and systems that the human mind and imagination cannot hope to master or comprehend, but for which we are nonetheless responsible.

Such is the technological aesthetic I perceive to be at work in McElroy's *Plus*, Pynchon's *Gravity's Rainbow*, Don DeLillo's later novels, and the one novel by Mailer—*Why Are We in Vietnam?*—that does not put style up against technology. When we contemplate technology in these works, we are ultimately contemplating ourselves, for the expressive potential within the machine returns us to the source of all thought, the human mind in relation to other minds. We may not be able to comprehend the abstract network of contemporary technological forces, as writers in the tradition of late nineteenth- and early twentieth-century naturalistic fiction believed they could represent and extrapolate from the forces contained in more visible steam, steel, and coal technologies. But this representational insufficiency does not prevent us from reflecting imaginatively—or even acting—on the technology of our own time. Its products, operations, and reproductive methods can still be brought together as an ad hoc, indeterminate sign system that does not require consistency among its parts.

Technology and a willing abstraction in these novels, as in the paintings of Marjorie Welish, allow the artist to achieve the coherence of vision that is necessary for any aesthetic reflection. Welish's technological imagination employs a visual system of geometric sectionings to suggest the networks and grids that underlie rational thought. But she does not attempt to resolve all visual contradictions in "a single total system which could dominate the whole of everything," as William S. Wilson writes in his essay on Welish's *Small Higher Valley* series. Welish sets two panels side by side, each with its own inner logic, although its columns, squares, and lines are structures flexible enough to accept the accident of a splash or paint-drip or submerged color showing through and clashing with the layers of paint above it. But the painting's most forceful expression occurs at a panel's edge or at the boundaries between panels, where wholeness is achieved by remaining open to differences. Not every part is or need be adapted to the needs of the whole, and the mind that perceives the whole need not lose itself in a romantic unity beyond alienations.

It should not be surprising that the machine metaphors of contemporary fiction are often dismissed as neurotic. As Haraway points out, to think that machines were "self-moving, self-designing, autonomous . . . was paranoid." Or used to be. But "now we are not so sure. . . . Our machines are disturbingly lively, and we ourselves frighteningly inert" ("Manifesto" 152). One of the more radical and surprising responses to this disturbing situation has been not to separate ourselves further from the machine, but to make technological mechanism the site of an aesthetic embodiment where mind and world are neither opposed nor merged (as they are for the paranoid self, which constantly risks isolation or loss of identity because everything in the world refers back to the self). Rather, all such binary terms can be integrated as separate elements in a single heterogeneous structure, most nearly approximated in current theories of literature and science by Haraway's "network" or "integrated circuit" on the one hand and Wilson's "field" on the other. In contemporary fiction we are witnessing increasingly frequent figurations of technology as an outward embodiment of thought, and this embodiment—whatever term we choose for it—would seem to have moved narrative away from romantic and modernist preoccupations with order, autonomy, and organic models of wholeness.

In recognizing both the necessity of and possibility for a bodily as well as a mental integration, Haraway, Wilson, Mailer, Pynchon, McElroy, and DeLillo are perhaps closer to one another than any one of them is to Adams; and each of these writers imagines material, historically situated realities that resemble one another much more than they resemble the simulated realities and virtual environments that technology in the era of cyberpunk fiction has aspired to produce. "Adams was still preoccupied with mind," writes John Christie, "Haraway preeminently with body," albeit a cyborg body "possessed of a scrupulous infidelity to origins, dislocated from the organic, the reproductive, and hence the Oedipal" (174–75). In Adams, thought and the imagination are the forms taken by human energies, although neither one is compatible with the "vast stores of new energy" that had been unleashed by turn-of-the-century technologies. This incompatibility, and the imagination's subsequent failure to represent itself to itself, are the inevitable results of a larger categorical discrepancy between a consciousness that uses words and ideas and an essentially nonverbal, and hence for Adams "supersensual," universe of force (Adams 381 passim.).

The postmodern writer would rely on images of embodiment to get in touch with this nonverbal reality, and to avoid ideals out of time that tend to be destructive to the world as it is. But these bodily integrations can be just as ironic, paradoxical, and programmatically incomplete as Adams's mental integration. A fascination with the body and its voluminous fragmentary histories emerges at the very moment when the body has ceased to be meaningful as a standard for measurement or symbolic comprehension. The functional elements in Haraway's "integrated circuit," for example, are not even visible, and its abstract location at the interface between neural and silicon chip networks tends to create in the user a sense of disembodiment. Not only did the bodybuilding, motorcycle transportation, and tattoo imagery in Acker's *Empire of the Senseless* anticipate and encourage these three fashion trends of the late 1980s and early 1990s; but also, like these trends, Acker's signature images are symptomatic of a tendency toward the experience of disembodiment in our ordinary lives. Whole stretches of the postmodern city are no longer even walkable, and so are unavailable for the realistic urbanism of a Balzac, Dickens, Dreiser, or Dos Passos, as well as for the charged symbolism of the modernist flaneur. Ever more frequently architects and engineers are coming to rely on computer simulations to reconstruct the sensual experience that is felt increasingly to be passing from ordinary life. And when the body is denied an integral if not necessarily an "organic" place in the ordinary lifeworld, it has a tendency to become itself the scene of writing and self-expression.

With these most typical of cyberpunk images (along with the "mirror shades" that allow the wearer to look out but nobody else to look in), have we not come full circle to Adams? In fact there is one well-known passage at the very end of "The Dynamo and the Virgin" that comes remarkably close to a cyborg image of embodiment. In this passage Adams presents not the body of the virgin, and not the dynamo mechanism, but writing itself as the material means—albeit a prosthetic one—of escaping what appears to be a purely intellectual quandary. Adams writes:

> The secret of education still hid itself somewhere behind ignorance, and one fumbled over it as feebly as ever. In such labyrinths, the staff is a force almost more necessary than the legs; the pen becomes a sort of blind-man's dog, to keep him from falling into the gutters. The pen works for itself, and acts like a hand, modelling the plastic material over and over again to the form that suits it best. . . . The result of a year's work depends more on what is struck out than on what is left in; on the sequence of the main lines of thought, than on their play or variety. (389)

Writers of the late twentieth century tend to be less inclined than Adams to see the conceptual universe as a "unity," even a paradoxical unity in the "multiplicity" experienced by the single imaginative consciousness. The postmodern writers I discuss in this book are certainly more accepting than Adams of "play or variety" in the "main lines" of their thought. Yet they share with Adams a self-consciousness that, though it often seems to lead them to an autonomous world of thought outside the body, is in fact deeply rooted in the materiality of contemporary forms of production, not the least of which are the forms of writing itself. In the unceasing craftedness of Adams's sentences, their simultaneous detachment from and immersion in American matter, their embrace of contradiction and self-consciously literary reflexings, we approach a properly complex relation between technology and the imagination. The high modernist's concern with the act of composition need not lead to a display of virtuosity for its own sake, satisfying the final criterion of a technological culture for the "best possible performance" (Lyotard, "Question" 77). Such virtuosity may also be the basis for a sharable, "intense pleasure in skill, machine skill," which Haraway lists as a necessary "aspect of embodiment" ("Manifesto" 180). The recurrent figure of the writer writing is more than a standard modernist topos, and more even than a metaphor for the technological construction of society as a whole. A technology in itself, the compositional act is yet another means—perhaps the primary means—for the postmodern imagination to represent itself to itself.

Here I concur with Jameson, who says in the preface to *Marxism and Form* that "any concrete description of a literary or philosophical phenomenon . . . has an ultimate obligation to come to terms with the shape of the individual sentences themselves, to give an account of their origin and formation" (xii). The writers discussed here certainly demonstrate his maxim that "real thought demands a descent into the materiality of language and a consent to time itself in the form of the sentence" (xiii). Such writing often produces *difficulties* which keep readers from understanding too quickly, and which, as a constant material irritation, stimulate readers to a perception of objects outside the grasp of a more directly "communicative" language. One finds such a frictional opaqueness in certain works of the Language poets, including Welish, who writes (after Robert Rauschenberg's figure of birds that have freed the stop signs) of "ruffled feathers exciting, agitating protocols, / oh, confusion!"

Jameson, however, is not content to rest in confusion. He is ideologically committed to resolving the signifying process in a dialectical totality, whereas the writers I consider, with the exception of Mailer, do not always conceive of the mind's relation to its own materiality as necessarily dialectical. Haraway, for example, prefers a "network ideological image, suggesting the profusion of spaces and identities and the permeability of boundaries in the personal body and in the body politic" ("Manifesto" 170). And Wilson develops a nondialectical, nonironic, and never completed "field of mutually supporting implications" ("And/Or" 25), in which terms such as *abstract* and *concrete*, *origin* and *formation*, *thought* and *materiality* impact on one another dynamically, without being leveled or subsumed in a single theoretical framework such as Jameson's Marxism. Leaving aside the likely incompatibilities between Wilson's earthbound field and Haraway's electronic network, the binary movement of thought in both would go beyond dialectic, with its implied longing for the resolution of all conceptual contradictions; their thought would even transcend metaphor, in McElroy's description of his own distinctive novelistic project, "and [work] toward homology," a side-by-side arrangement of separate discourses (LeClair and McCaffery, *Interviews* 242).

Don DeLillo's description of his novel *Ratner's Star* as "a piece of mathematics" (LeClair and McCaffery, *Interviews* 86), like McElroy's conception of *Lookout Cartridge* as a "computer in itself" (LeClair and McCaffery, *Interviews* 244), are only the most obvious expressions of a commonality of purpose, typifying a habit these writers have of using technical terminology to justify their work's supposed difficulty. Such conceits could also derive from the training or practitioner's experience many of these writers had in technical fields: Pynchon majored for a time in engineering physics at Cornell University and Mailer in aerodynamics at Harvard; DeLillo spent his apprentice years in advertising and "freelance writing, some of which was technical" (LeClair, *In the Loop* 16). But the affinities that unite science, technology, and literature run deeper than the professional circumstances of one particular grouping of writers, and these affinities are revealed most significantly, I believe, in the compositional struggle itself, even in "the origin and formation," to borrow Jameson's term, of individual sentences.

The methods of observation and organization in literature, its conceptual processes and formal arrangements ("modelling the plastic material over and over again to the form that suits it best" are of a piece with technological modes of production. Even the process of literary self-creation, typically defined as the romantic expression of autonomy and personal freedom, is no less interesting when it parallels technological processes that are thought to oppose or oppress the self. Such a compositional activity involves, in Walter Benjamin's formulation, "a mimetic, not an instrumental skill," a matching of technology with the material of language that Benjamin calls the "central intellectual task of the modern era" (Buck-Morss 70, citing Benjamin's *Arcades Project*). At a time when those working in music or the visual arts might seem to be closer to practical technological operations, I hope to show that literature is still able to reveal the "new significance of artists, who are socially useful precisely as experimenters who discover the human and cultural potential within the new technology" (Buck-Morss 30). I do not, however, regard the compositional process as a way of reclaiming for literature the centrality and professional status of the sciences. I am most interested in those narrative moments when a literary figuration *fails* to match its technological object, since this is the point at which literature can begin to represent not technology itself but the tumultuous and incongruous nature of postmodern experience. The two discourses of literature and technology *should* remain separate in order to facilitate real communication and critical interchange between them, just as the writer should sustain the position of a cultural outsider in order to work most forcefully and eclectically within society and against the all-compelling hold of its ideology.

That the writer remains marginal and resistant to the technological culture is not in itself news; indeed, the predominant postmodernist concern with the compositional self preserves a romantic and modernist impulse to stand apart, to get outside the space of technological production and cultivate an aesthetic detachment. As Jameson among many others has pointed out, however, there is no longer any "outside," no Archimedean vantage point from which to base a critique of the technological culture. For the postmodern writer exclusion is not a prerequisite to seeing technology and society whole, nor is it simply any longer a reflection of the writer's metaphysical (read: socioeconomic) marginalization before vast formations of otherness (read: corporate capital). More likely exclusion would seem to be a tactic whereby the writer can maintain contact with the "ordinary conditions of life" that are so necessary—as Tolstoy remarked in *Anna Karenina*—for forming a conception of anything (830). In themselves, such conditions are never sublime, but it is only through them that sublime emotions—Nikolay's grief and his brother Levin's great domestic joy—can be felt (831). And this is no less true when domesticity itself is technologized and made to exist amid the white noise of televisions and supermarkets.

The quest for the ordinary in postmodern American literature is no more or less evident in writers as private as Pynchon, McElroy, DeLillo, and Gaddis than it is in Mailer, who acts out the gaudy transformations and metaphorical linkages of a private imagination in full view of the national media. DeLillo fashions himself an "outsider in this society" (DeCurtis 281); Mailer characteristically adopts an all too knowing insider stance. The life-style choice is theirs. As writers who publish their books through large corporations, however, they are equally within the sublime ideology that holds them, equally dependent on it to give shape and dimension to the unredeemed particulars of ordinary life, but always "constructing, abiding, and challenging such sublimity from within," as Rob Wilson says of their counterparts in postmodern American poetry (11).

The desire of the literary outsider to "enter history," though an implicit theme throughout this book, becomes of increasing importance in the later chapters on Don DeLillo, where the late modernist "death of the author" is presented as a defining characteristic not of the nonreferential *écriture* celebrated by Roland Barthes in his classic 1968 essay of that title, but of a more recent development in contemporary American writing which I call here a postmodern or conceptual naturalism. Unlike traditional naturalism, such writing has not depended on an appropriated "truth" of contemporary science—Dreiser's Darwinism, Zola's thermodynamics, Norris's Lombrosian criminology, Adams's evolutionary history, or Dos Passos's relativity—to impose order and meaning on the welter of contemporary phenomena. It does not focus on matter to the exclusion of mind. Nor has it depended, as Tom Wolfe would have the novel depend, on the writer's perfection of "research" methods to make successful use of the journalistic material that a technological culture produces. The postmodern writer is not trying to compete with the scientist or the journalist by claiming an objectivity and authorial impersonality that would rise above history—although Mailer embraces a welcome, if suspect, objectivity in *The Executioner's Song* and *Harlot's Ghost*. Rather, like Benjamin's angel of history, the contemporary naturalist writer disappears into the wreckage of everyday culture, wherein the culture might find its own direction against the continuing storm of a progressivist history.

In its detailed attention to the "ready language" that can exist outside any literary framework (Buck-Morss 17, citing Benjamin, *One Way Street*), and in its crafted style and constant reference to the material world as a medium or shifting ground for philosophical reflection, the writing of Walter Benjamin provides a strong modernist model for contemporary narrative constructions of history. But no sooner do we approach the details of a postmodern culture than the contradictions in speaking of a "naturalist" historical construction become evident. The "aura" surrounding DeLillo's "Most Photographed Barn in America" clearly is not the aura of Benjamin's "Work of Art in the Age of Mechanical Reproduction," for DeLillo's aura is "sustained," not degraded, by collective processes of technological reproduction (*White Noise* 12). Whereas aura for Benjamin is another myth of lost origins, referring to works of art whose authority depended on the specific place of production and was "tied to [the] presence" of a performing self (229), the characters in *White Noise*, *Libra*, and *Mao II* have their existence in a continuing present and in the "place" described by Haraway as the "scary new networks" making up the "informatics of domination" ("Manifesto" 161). Benjamin's understanding of nostalgia for the past is reversed, in DeLillo, to become a nostalgia for one's own present and a feeling of missing a place when one is already there: "We're here, we're now," enthuses one of DeLillo's characters on beholding the barn (12–13). Indeed, any novelist capable of imagining the SIMUVAC operation, in which actual disasters are welcomed as occasions to perfect a model future evacuation, might seem to have more in common with Jean Baudrillard and his blithe contemporaneous history of the "simulacrum culture" than with Benjamin and his backward-looking dialectical materialism.

A whole world of simulations opens up in the narratives under consideration here—in McElroy's part mechanical, part vegetable, computerized but never "artificial" intelligence; in the flight simulations that NASA astronauts underwent in training for an actual flight to the moon that was designed at every point—as Mailer noted—to be *broadcast;* and in Pynchon's Disneyfied tour of the underground Mittelwerke factory where the V-2 rocket was at last produced and sent into the field of wartime operations. Pynchon, haunted by the "ghosts" that the historical rocket created, rightly resists the ahistoricity (though he is not without the impish perversity) that Baudrillard's commitment to a simulation culture can encourage (see Christopher Norris on two Baudrillard essays that counsel readers not to believe in the reality of the Persian Gulf war). But when the narrator of *Gravity's Rainbow* says that the ghosts' "likenesses will not serve" (303), this narrator refers to (and for the moment is aligned with) an older order of simulation, an order in which appearance and reality could be rationally distinguished, and when human and machine were "counterfeits" of each other joined only by analogical "*resemblance*." In an age of robotics and electronic miniaturization, the machine has absorbed all appearances, it has become "man's *equivalent* and annexes him to itself in the unity of its operational processes" (*Simulations* 178). In the order of electronic simulation described by Baudrillard, the older styles of critical resistance (even those in *Gravity's Rainbow*) lose their meaning.

The absorption of appearances into the machine is consistent with the present argument that an immanent meaning is to be found in technology's own constructions. But I would also argue, contra Baudrillard, that the omnipresence of such simulations does not preclude a naturalistic construction. The machine is not "man's equivalent," and the loss of an analogical bridge for uniting conceptual dualities that are themselves suspect in the first place, instead of frustrating representation, might free writers to imagine styles of resistance that are not merely *oppositional*. By being attentive to the modes of technological simulation and refusing to oppose them with some purportedly "authentic" literary consciousness, writers can hope instead to locate sublime, nonobservable relations within technology itself, and so to discover within its systems "secrets" (to stay with DeLillo's terminology), which are themselves "almost secrets of consciousness, or ways in which consciousness is replicated in the natural world" (DeCurtis 299). The detailed engagement in actual technological simulations, if you will, might then provide an alternative to the potential passivity of Baudrillard's vision, an operational reality that is not simply a copy of a copy in a series that has no original but a naturalistic reconstruction that would make the simulacrum culture livable.

The readings that follow accept Lyotard's elevation of the notion of the unpresentable; they even accept Baudrillard's description of the technological culture as mediated through and through. But neither description is a reason for rejecting truth claims in the political and metaphysical realms or, in the aesthetic realm, for denying the power of narrative to transport us out of ourselves. I would not want to advance a postmodern equivalent to the great writing that Longinus spoke of, written by "godlike" authors who aim at the highest flights of composition "and overlook precision in every detail" (47). A respect for the facticity of postmodern reality—a reality outside the mind of the artist or historian to which people can respond—saves the four main writers in this study from linguistic solipsism on the one hand and, on the other, from the total relativism that more cynical pragmatists than Lyotard are prone to fall into. Imagination and fantasy in this fiction, at its best, are on the side of the real, to borrow Lacan's concise formulation. They model and participate in the unspoken desires that structure our social and material reality. And if the postmodern reality should turn out to be incomplete and undecidable, the loss of certainty can also produce "a modest gain in realism about one's powers" (Wilson, "And/Or" 19).

Modesty, of course, is hardly the first characteristic of the sublime. And neither is Mailer, the first writer I consider, especially known for his modesty. But before turning to some of contemporary literature's spectacular instances of sublime expression, including those by Mailer himself, I would like to begin with a close look at a deceptively modest performance, his primarily journalistic account of the flight of *Apollo 11*. *Of a Fire on the Moon* demonstrates that, even in sublime depths of uncertainty when "questions would only open into deeper questions" (90), there is no reason to abrogate the intellectual's power simply to seek out the best documentary evidence the technological culture provides (but most often imperfectly contextualizes and/or comprehends). To be part of the technological structure implies a validity that is no less real because the structure lacks a center and a foundation. Technology and imagination construct each other; their relation is complementary, mutually inclusive, and of necessity incomplete, like the famous suspension of history in the novel and of the novel in history in *The Armies of the Night*. The failed and rejected attempt to transcend history in *Fire* will begin to show what it means to construct a fiction that is integral, though never "adequate," to the technological production of reality.

1

Mailer's Psychology of Machines: *Of a Fire on the Moon*

There is early in *Of a Fire on the Moon* a moment that indicates how we should read Mailer, how he locates and dramatizes himself, and, more generally, how the romantic imagination of the late twentieth century engages with technology. Entering the massive Vehicle Assembly Building (VAB) at Cape Kennedy where the *Apollo* rockets were assembled, Mailer is immediately inside a structure whose interlocking patterns and open, shifting dimensions seem the very embodiment of technological complexity, suggesting networks of power and corporate control beyond the comprehension of any single mind or imagination. Standing inside "the first cathedral of the age of technology" and looking down at the floor "as much as forty stories below" (55), Mailer experiences a feeling of disembodied intellect and "loss of ego" (3), less a spiritual transcendence than a grudging and half-humorous confession of irrelevance in an increasingly secular and nonverbal world:

> All the signs leading to the Vehicle Assembly Building said VAB. VAB—it could be the name of a drink or a deodorant, or it could be suds for the washer. But it was not a name for this warehouse of the Gods. The great churches of a religious age had names. The Alhambra, Santa Sophia, Mont-Saint-Michel, Chartres, Westminster Abbey, Notre Dame. Now: VAB. Nothing fit anything any longer. The art of communication had become the mechanical function, and the machine was the work of art. What a fall for the ego of the artist. What a climb to capture the language again! . . . Yes, one would have to create a psychology to comprehend the astronaut. (55–56)

Mailer's characteristic response to technology—a source of new objects and literary images but, more important, a rationalized, reduced, and potentially dominant way of looking at the world—is to oppose its impersonal operation with an individual style, to recapture "the language" and create a form adequate not only to the experience of spaceflight but also to the entire system of multinational corporate production that made spaceflight possible. This literary response is more than a celebration of the alienated artist figure cut off from an inhospitable mechanistic culture. The threat of ego destruction stimulates the ambitious writer to build himself back into the technological structure that excludes him; and Mailer, who perhaps more than any literary contemporary has advertised his career as a romantic project for the self, finds his greatest challenge here.

At first glance Mailer's narrator appears subdued and even passive in the face of the accomplishment represented by the rocket and the structure that houses it. Feeling "like a spirit of some just-consumed essence of the past" (6), he regards his subject and the corporate structure from perhaps the only "outside" perspective he can imagine—that of a person who is already dead. Although such a perspective may seem incidental to Mailer's concern journalistically to "make an approach to the astronauts" (18), it is fundamental to his attempt literally to embody, in his own mind and senses, an emerging technological condition. Like the imagined communications from a psychic world "beyond the Zero" in Pynchon's *Gravity's Rainbow*, like the extraterrestrial growth of a "Body Brain" in McElroy's *Plus* and the "constructs" of once-living personalities in William Gibson's science fiction, and even to a degree anticipating the funereal perspectives in Mailer's own *Ancient Evenings* and *Harlot's Ghost*, the ghostly perspective at the start of *Fire* is intended at once to comprehend and contrast with the accelerated but never certain changes within contemporary American society. Each of these writers, by carrying perception beyond the point where rational thought can go, would put the structure of modern technology in a wider context—although for Mailer more than the others, such a context remains centered in the self, in the about-to-be-embodied consciousness that continues to trust in the authority of the senses to get in touch with the world.

Mailer tells himself that he must not "dominate this experience" intellectually, that he must "look instead to receive its most secret voice" (56). Yet despite his apparent receptivity there is much to suggest that he had worked out a "psychology" for the astronaut years before he ever thought to look into the details of space technology. He opens *Fire* with an account of the death of Ernest Hemingway, long a touchstone of his own ambitions not simply for celebrity but for embodying the age in his writing. While Hemingway was living, he "constituted the walls of the fort," making it possible to believe that courage and personal style (elsewhere regarded as the insulating layers of the masculine ego)[1] might yet enable one to live with dread and to retain at least a coherent personal identity in an age that lacked any dominant principle or organizing center. The suicide of America's "greatest living romantic," happening near the start of the decade that would end in the flight of *Apollo 11*, had left a void in the culture: "Technology would fill the pause" (4).

Mailer's own sentiments, it is clear, incline naturally to romanticism, or at least to the postromantic myth of the embattled ego that he himself had begun to develop more than ten years earlier in the ground-breaking essay "The White Negro" (1957). There he had made his earliest sustained attempt to elaborate a comprehensive social psychology, arguing that because our age is confronted with the triple threat of "instant death by atomic war," the memory of the concentration camps, and a stifling "slow death by conformity . . . the only life-giving answer is to accept the terms of death, to live with death as immediate danger, to divorce oneself from society, to exist without roots, to set out on that uncharted journey into the rebellious imperatives of the self" (*Advertisements* 339).

By thus insisting on the primacy of the self and, even more, by trying to preserve some private sense of his own mortality against the many forms of collective death implemented by the totalitarian state, Mailer could attack the "partially totalitarian," predominantly conformist society of the Eisenhower years. As early as *The Naked and the Dead*, Mailer had foreseen the possibility that wartime technology and managerial techniques that had "outraced the psyche" might be applied to the creation of a vast social machine, an incipient totalitarianism of the right that would produce irrational reactions and extremes of self-assertion in the resisting individual: "In the marginal area, the gap,

1. In *An American Dream* Mailer writes: "The spirits of the food and drink I had ingested wrenched out of my belly and upper gut, leaving me in raw Being, there were clefts and rents which cut like geological faults right through all the lead and concrete and kapok and leather of my ego, that mutilated piece of insulation, I could feel my Being . . ." (11–12). Not coincidentally, the subject here, as in the Hemingway passage in *Fire*, is the breakdown of masculine ego that occurs at the point of suicide.

were the peculiar tensions that birthed the dream" (391). With a radical psychology and sociopolitical rhetoric largely taken over from Wilhelm Reich, Mailer in the late fifties was imaginatively occupying that gap, immersing himself in the psychopathic consciousness of the "hipster" or "white Negro" who was capable of expressing those irrational desires, class truths, and psychic contradictions that a technocratic culture generally suppresses or officially ignores. Because the hipster was, in Mailer's view, trying to free himself from an imprisoning and historically antiquated nervous system that carried in the style of its circuits "the very contradictions of our parents and our early milieu" (345), he was suited to express the contradictions of an entire society. In the absence of any immediate correspondence between private experience and the facts of communal reality in America, it was Mailer's faith that the psychopath could still achieve psychological stability through an inner "communication of art," and so overcome the separation from experience that a technologized society purportedly produces (341).

In *American Fiction in the Cold War* Thomas Schaub usefully contextualizes this heady argument, with all of its "radical" implications for self-liberation, as a pure product of contemporary liberalism. The valorization of personal and psychological experience in postwar writers as various as Mailer, Flannery O'Connor, Saul Bellow, Ralph Ellison, John Barth, and Jack Kerouac is described as "a romantic regress, one that claimed authority for the radical integrity and idiosyncrasy of the artistic consciousness" (Foreman 182). The postwar exploration of a personal psychology—and the novelist's correspondingly frequent adoption of a first-person narrative, modulated to a self-watching third person in Mailer's sixties journalism—in retrospect might be understood as far less radical than it seemed at the time. For Schaub's argument is not only provocative (Mailer a liberal!) but convincing in the wealth of contemporary political discourse that is brought to bear on Mailer's essay. Schaub shows that it was quite common among mainstream liberals and conservative intellectuals of the time to conceive human society as a mental structure, and to value individual creativity as the last remaining means of resisting a society perceived to be increasingly uniform and "totalitarian" (this last a liberal internalization of conservative anti-Soviet rhetoric). Indeed, nowhere is Mailer more consistent with mainstream consensus ideology than when he is resolving the psychic contradictions of "the white Negro" in a cult of creativity and unconstrained desire—the apotheosis of liberal individualism.

One area, however, in which Mailer can be said honestly to resist both consensus liberalism and the fifties culture of psychoanalysis is in the essay's presentation of sense experience, the loss of which Mailer directly relates to the increased technologization of contemporary life. The hipster's search in the nightlife of the modern city for physical danger and sexual adventure was less a celebration of individuality than a struggle for psychic wholeness and physical embodiment. And the two sensations fundamental to this struggle—lust and anxiety—were less a means of unbridled self-expression or panicked retreat from the prospect of nuclear destruction than they were ways of presenting experience in its felt immediacy.[2] Extremes of lust and anxiety, and a psychopathic consciousness receptive to spirits unseen by a rationalized consensus culture, become the condition of the Mailer hero, for in none of these experiences is the hero capable of separating himself from what is happening to him. Taken outside himself, he is made to enter an "enormous present" in which individual actions have experiential significance and direct metaphorical connection to people and events around him, rather than being perceived through the rational categories and technological media that distance us from our minds and bodies.

Hence, for all the violent self-assertion of the earlier work, the loss of ego that Mailer experiences at the start of *Fire* is not new for him. What is new, though, is the situation of the self in the very midst of a structure in which "nothing" human "fit anything any longer" (56). Mailer's vertigo above the Vehicle Assembly Building is "the vertiginous feeling," in Peter Brooks's phrase, "of standing over the abyss created when the necessary center of things has been evacuated and dispersed" (21). For Mailer, however, any attempt to reconstruct the lost center must be grounded in human drama and in metaphor, in what could be transferred out of the technological context he encountered at the Manned Spacecraft Center into an expanded imaginative context.

Through Steven Rojack, professor of existential psychology and hero

2. In his book *The Reenchantment of the World,* the historian of consciousness Morris Berman follows Reich in expressing the loss of self that may occur in sex and in moments of panic: "As I make love to my partner, as I immerse myself in her body, I become increasingly 'lost.' At the moment of orgasm, I *am* the act; there is no longer an 'I' who experiences it. Panic has a similar momentum, for if sufficiently terrified I cannot separate myself from what is happening to me" (76). Two experiences, the sexual and the psychotic, are thus central in Mailer because they resist rational control and return us to a direct bodily apprehension of the world bordering on the mystical.

of *An American Dream,* Mailer had already proposed "the not inconsiderable thesis that magic, dread, and the perception of death were the roots of motivation" (8). But in no other work does Mailer explore Rojack's thesis in so detailed and systematic a fashion, and if the predominantly journalistic form of *Fire* prevented him from realizing the many imaginative possibilities of the original draft, the concerns of these chapters persist. *The Alpha Bravo Universe* became the working title of a massive fiction Mailer conceived during the seventies, a three-part novel whose first part, itself over seven hundred pages long, was to become *Ancient Evenings.* This trilogy appears to have been abandoned—as have nearly all of Mailer's projected novels from the time of *Advertisements*—in favor of a much different book, a "novel of the CIA" called *Harlot's Ghost,* whose first part, "From the Alpha Manuscript," is nearly twice the length of *Evenings.* In the researches of one character, Hadley Kittridge Montague, the Alpha and Bravo psychology is taken over and more or less worked out (they are called Alpha and Omega in the book). Thus, the dream narrative persists, whether or not Mailer completes this project, and the "Alpha and Bravo" manuscript survives as its earliest, most explicit articulation.

In a writing life that is often regarded as fragmentary, opportunistic, and hedged in by contingencies of the moment, the "Alpha and Bravo" project represents a surprising continuity, to the point of obsession. That this project has approached a final form should alone necessitate a revisionary reading of Mailer, one that places the concerns of *Fire* closer to the center of his ouevre. In this reading the difficult relation between technology and the romantic imagination can be seen not as an isolated moment in Mailer's development but as a concern that is integral to his largest ambitions and thus fateful as well for the fast-disappearing romantic self in contemporary American writing.

The Romantic Background

Introducing an important collection of essays on Mailer, Michael Lennon rightly resists the tendency of much contemporary criticism to focus on Mailer's biography. Lennon presents Mailer first as a "connoisseur of narrative forms, styles, and perspectives" whose "constant self-admonition has not been simply, as often claimed, to find ways to advertise himself, but also to see himself," in Mailer's words in an interview, "as a piece of material, as a piece of yard goods. I'd say, 'where am I going to cut myself?' It's a way of getting a psychoanalysis, I think" (Lennon 2, 3). The self as a material or medium to be publicly manipulated and a consciousness separate and impersonal as a "conveyor belt" (as he described himself in *The Executioner's Song*) are the two poles of Mailer's complex narrative presence (Lennon 3). These opposing tendencies toward self-dramatization and detachment can be found both within particular works and throughout his career. Yet it is also significant, I think, that in each case Mailer uses a technological metaphor to describe himself. Much of Mailer's celebrity performance is the spectacle of a personality constantly forming and re-forming itself in an effort to capture the accelerated changes within a technological society, and however much he may oppose the media and other corporate mechanisms of that society, his involvement with them is inevitable. His need to create a new style "for every project" (*Armies* 35) and his frequent compulsion to work under deadline (which he celebrates as a professional virtue) would be possible only in a society that demands free enterprise and continued technical innovation, while his ambition to represent American social reality in its entirety depends on technology, in the form of production services and vast communications networks, as at least a plausible figure for the invisible operation of corporate power in the world.

The narrator of *Fire* has a presence that is somewhere between the apparently disembodied narrator of *Song* and the autobiographical personae of the two "nonfiction novels" that preceded *Fire: The Armies of the Night* and *Miami and the Siege of Chicago*. Implicit in all of Mailer's self-characterizations during the sixties was the idea that he could best discern and express the social upheavals and political conflicts of the time by becoming "an egotist of the most startling misproportions, outrageously and often unhappily self-assertive," one who could, if not quite comprehend the age, at least exemplify its larger pattern (*Armies* 66). Now in *Fire* Mailer, "detached this season from the imperial demands of his ego" (6), portrays himself modestly as "little more than a decent spirit, somewhat shunted to the side" (4), using the name Aquarius but knowing that, as America moved into the seventies and the celebrated Age of Aquarius, he had "never had less sense of possessing the age" (4). His identity unanchored amidst technological forces that seem altogether antithetical, faceless, and impersonal, Mailer finds himself in the position of Henry Adams confronting the dynamo in the Great Exhibition of 1900: contemplating the moon shot,

Mailer can no longer embody or give shape to the social and historical reality.

His compositional difficulties are real enough, but one should not be taken in by his modesty; like Adams's assumed modesty in the *Education*, it has its rhetorical uses. Mailer had already evoked Adams in *The Armies of the Night*, his first sustained attempt to speak of himself in the third person, and his subject and mood in *Fire* are even more in line with Adams's self-deprecating irony. For Adams, writing at the dawn of twentieth-century modernity, third-person narrative was a way to regain his own place and imaginative coherence in a universe of forces that tended to exclude him. It was a way, that is, both to represent and to participate in the cosmos, and to maintain a radical freedom by being at once inside and outside an intellectual system.

Freedom is very much the issue in the chapter of *Fire* titled "The Psychology of Machines," in which Mailer absents himself as a personality only to project human qualities onto the rocket. Here Mailer is interested in studying primarily those aspects of the machine that are most prone to breakdown, choosing to interpret malfunctions as evidence not of operator error, design inadequacy, randomness, or mere entropic decline but of a conceivable psychology inhabiting the inert mechanism. At first glance this psychologizing of the machine may appear little more than a way of enlivening an otherwise straight technical narrative of space rocketry (and "The Psychology of Machines" is easily the most entertaining chapter in "Apollo"). But in the light of Mailer's polemical intentions, the theoretical argument has a greater urgency, amounting to an almost desperate assertion of human freedom against the threat of total cultural absorption into the machine. To counter the scientific bureaucrat's supposed reliance on technology to insulate the self from dread, Mailer would introduce anxiety into the machine itself; for every physical cause the technologist might produce to explain mechanical failure, Mailer postulates prior causes beyond the domain of physics. Advancing his ideas in accordance with received conventions of rational argument, he locates his speculative psychology at the point where its existence ceases to be arguable:

> For every malfunction there is a clear cause technology must argue, a nonpsychological cause: psychology assumes free will. A human being totally determined is a machine. Psychology is then a study of

> the style of choice provided there is freedom to choose. Even a title like The Psychology of Machines assumes that the engine under study, no matter how completely fitted into the world of cause and effect, still has some all but undetectable horizon between twilight and evening where it is free to express itself, free to act in contradiction to its logic and its gears, free to jump out of the track of cause and effect. (162)

Mailer is ready to admit that his psychology lacks a scientific basis, but how could it not? "Since such events take place, if they do take place, on those unexpected occasions when no instruments are ready to examine the malfunction, the question is moot. No one alive can state to a certainty that a psychology of machines exists or does not exist" (162). The very fact of breakdown, its persistence in the face of all social, scientific, and technological determinations, is for Mailer solid enough ground on which to build a psychology of machines, one that could be said, however improbably, to apply the ethic of nonconforming autonomy from "The White Negro" to the malfunctioning or conceivable self-functioning of the machine.

Again, for all the "zany theoretical seriousness" of the argument (Poirier, *Mailer* 163), Mailer is not the first writer to have thought this way. It was one of Samuel Butler's "Professors of Inconsistency and Evasion" who had argued, in "The Book of the Machines" chapters of *Erewhon*, that spontaneous breakdowns might be taken as proof of a machine's vitality or of some vital spirit manifesting itself through the machine—what we call "spontaneity" being, according to Butler's Erewhonian professor, "only a term for man's ignorance of the gods" (219). Anticipating Mailer's conviction that "dread inhabited the technology of rockets" (*Fire* 169), Butler endows the machine with a consciousness and "a will of its own," meaning that, however much we believe we can understand and control the forces the machine sets in motion, there always remains something "Unknown and Unknowable" about them (*Erewhon* 215; Mailer stresses and repeats the phrase "a mind of its own" in *Fire* 162).

Regarded this way, Mailer's psychology can be seen as an assertion—part political, part religious—of what Thomas Pynchon would later describe as "our faith in Malfunction as still something beyond Their grasp," the persistence of freedom and contingency despite corporate efforts to dominate and restrain (*Gravity's Rainbow* 586). But that interpretation makes Mailer sound more helpless than he is in the face of NASA's power. For even as he steps outside the agreed-upon modes of rational argument and scientific debate (where a hypothesis cannot be validated unless it is subject to proof by experiment), he opens the discussion to a whole range of irrational human responses that science is obliged to ignore, and these may include the responses of scientists, engineers, and astronauts as well as those of the laity. Rational argument itself, from his viewpoint, would seem to be a linguistic machine built of interlocking logical components that is unable to accommodate the mystery and nuance inherent in experience, technological or otherwise. Only by taking an imaginative leap or "flight" into metaphor, Mailer seems to imply, can we hope to understand the human dimensions of the flight to the moon.

It is therefore as an expression of our anxiety in the presence of the unknown that his psychology has its chief relevance. For what better way than through a psychology of machines to communicate the tensed activity aboard the lunar module and among the Mission Control engineers during the last four minutes of descent to the moon? It is here, in the brilliantly narrated chapter "The Ride Down," that Mailer's psychology has perhaps its finest application, helping to create for the reader something of the theater that developed when, during the last lunar orbit before the final decision to risk a landing, radio communications began inexplicably to drop out and the display panel flashed the alarm "1202," signaling an information overload in the on-board computer (a development that could have been fatal to the mission).

It is, as George Landow has perceived (*Jeremiahs* 185), an existential moment, one that demands more of the astronauts, guidance officers, and Mission Control technicians than the machine alone can give. And yet the feeling conveyed by Mailer's account of the landing is less one of momentary freedom and openness to chance than that of a greater constriction of will and attention within the binary determinations of the machine. The limits of decision at every stage of the module's flight, GO or NO GO for the mission's continuation, were in place weeks before the warning signal appeared; the navigation and guidance officers were already equipped with established procedures and a clear chain of responsibilities for handling the crisis. All through the landing flight controllers, like mechanical extensions, are "screwed to the parameters of the consoles," and the astronauts "come down toward the gray wife of the earth's ages with their eyes riveted to the instruments" (379). Anxiety, an experience of the body and the psyche, resides in the smallest unforeseen occurrence or hint of malfunction, so geared to the machine are the thoughts and emotions of all those involved in the flight.[3]

Mailer concludes "The Ride Down" with a final reference to the imaginative reaches of the technologist's consciousness:

> And Kranz, who had issued every order to [flight controller Charles] Duke, and queried his controllers in a voice of absolute calm for the entire trip down, now tried to speak and could not. And tried to speak, and again could not, and finally could unlock his lungs only by smashing his hand on a console so hard his bones were bruised for days. But then if his throat had constricted and his lungs locked, his heart stopped, he would have been a man who died at the maximum of his moments on earth and what a spring might then have delivered him to the first explorers of the moon. Perhaps it is the function of the dream to teach us those moments when we are GO or NO GO for the maximum thrust into death. They were down, they were on the moon ground, and who could speak? (381)

Readers familiar with Mailer's earlier work will realize that the conceit of the moon as an angel of death originated not in any immediate involvement with the events of the control room but in Mailer's prior psychological representations in his fiction. In the first of many such moments in *An American Dream*, for example, Steven Rojack, standing with both feet over the balustrade of a tenth-floor apartment in Manhattan, feels the moon urging him to jump: "My body would drop like a sack, down with it, bag of clothes, bones, and all, but I would rise, the part of me which spoke and thought and had its glimpses of

3. A passage in *Gravity's Rainbow* makes explicit the feelings that may arise whenever one's fate is given over wholly to the inert mechanisms of technology. Enzian, leader of the black Southwest African commando unit devoted to assembling and firing a rocket modeled after the German V-2 (originally designated Aggregat 4 or A-4), describes the engineer's bodily and psychic identification with the rocket: "One reason we grew so close to the Rocket, I think, was this sharp awareness of how contingent, like ourselves, the Aggregat 4 could be—how at the mercy of small things . . . dust that gets in a timer and breaks electrical contact . . . a film of grease you can't even see, oil from a touch of human fingers, left inside a liquid-oxygen valve, flaring up soon as the stuff hits and setting the whole thing off—I've seen that happen . . . rain that swells the bushings in the servos or leaks into a switch: corrosion, a short, a signal grounded out, Brennschluss too soon, and what was alive is only an Aggregat again, an Aggregat of pieces of dead matter, no longer anything that can move, or that has a Destiny with a shape" (362; Pynchon's ellipses).

the landscape of my Being, would soar, would rise, would leap the miles of darkness to that moon." Rojack, no less than Kranz, is held immobile within a set of binary determinations. The voices he hears as though from separate compartments of his psyche send contradictory commands ("Instinct was telling me to die. . . . Which instinct and where?"), and as he releases one hand from the rail, the other tightens its grip (12).

Even though Rojack chooses not to take the drop, he does come afterwards to feel "as if I had died and did not altogether know it, [the way it might be] for the first hour of death if you chose to die in bed—you could blunder through some endless repetition believing your life was still here" (14). (So too would Menenhetet IV experience the first moments of death at the opening of *Ancient Evenings,* moving toward "some existence on the other side" until he has "a body again" [4, 5]). This is how Mailer would like to view the many habitual activities and ways that make up the working life of the astronaut and the engineer: their routinization of language to match the stripped-down elegance of machinery; their apolitical acceptance of mechanisms of corporate control, be they sociological, procedural, or technical; their immersion in work that insulates them from the present and their ability to live with comfort in the future; their discipline, their thrift, and—what is at the heart of Mailer's creation of a psychology of machines—their preoccupation not with nature itself but with the endless balance/imbalance of its technological simulations. In *Of a Fire on the Moon* repeated and endless approximations to the real are experienced by the astronauts and NASA technicians as if these men were in a psychic limbo, an undifferentiated repetitive existence not unlike the one Rojack feels coming over him like a malignancy. And if Rojack in the course of the novel must make himself over again to free himself from these stifling conditions, Mailer seeks an analogous regenerative power in the drama of the landing.

That he had fallen short of this ambition can be felt, however, in the reticence with which these late speculations are presented, suggesting that Mailer could not finally bring himself to insist on his own prior metaphorical flights. His largest claims for the dream continue to be couched in the subjunctive mood ("*Perhaps* it is the function of the dream . . ." [381; emphasis added]); and he does little with the idea of a psychic journey after his first conjecture about the astronauts' secret belief that "the gamble of a trip to the moon and back again, if carried off in all success, might give thrust for some transpostmortal insertion to the stars. Varoom!" (35). Such last-minute reflections, recalling as they do Mailer's earlier thoughts on interpsychic communication, creative orgasm, embodiment, and the perception of death, presume a fictional development that *Fire* never realized. Still less is Mailer able here to extrapolate an inclusive social psychology from the psychology of astronauts. That he had not even made an approach to the high society within NASA, "a group as closed to superficial penetration as a guild of Dutch burghers in the Seventeenth Century," had been the subject of his meditations just a few pages before the account of the moon landing: "No one but the men in that room would ever begin to know the novels and dramas of conflict, the games of loyalty, and what captures and frustrations of power had played back and forth among these men in the last ten years—it was another of the great novels of the world which would never be written. And was the world a little more polluted for that?" (366).

What Mailer achieves, in chapters such as "The Psychology of Machines" and "The Ride Down," is a compelling meditation on technology's existential dimensions, its shaping influence on the ways in which we perceive reality and imagine for ourselves an authentic personal identity. But beyond this achievement, unmatched in the journalistic accounts and histories, there is a particular "great novel" of American technology that Mailer abandoned to the discarded drafts of "The Psychology of Machines." These drafts represent his unprecedented attempt, more ambitious in conception than what survives in the published text, to rid himself of the standard liberal-intellectual objections to the space program, to pare down his own personality, and to internalize the objects and methods of space technology. In the drafts we can more closely follow Mailer's attempt to narrate the stratifications and lineaments of American society and begin to distinguish more sharply between possibilities that are still available to the romantic ego and those that no longer seem to be.

The Alpha and Bravo Universe: Early Drafts for "The Psychology of Machines"

In the warehouse on Sixty-second Street in Manhattan where Mailer's papers are kept, six large boxes contain a late draft and source material for *Of a Fire on the Moon*. One of these boxes holds a half dozen science texts, issues dating from July 1969 of the *New York Times Magazine* and the daily *Times,* lightly annotated copies of *Science* (July 20, 1970) and *Scientific American* (May 1968), "more than one technical manual" (Mailer's understatement [*Fire* 152]), hundreds of photocopied NASA press releases and flight transcripts, and some forty pages of double-spaced draft labeled "ALPHA AND BRAVO. Section written for OF A FIRE ON THE MOON . . . not included in the final text of the book." Essentially a meditation on the "grace and economy of communication" between the two halves of a divided psyche, "mythical Alpha and Bravo" (57), this section dovetails quite coherently with the published text. One key passage, for example, on the purported split personalities of the astronauts Neil Armstrong, "Buzz" Aldrin, and Michael Collins recurs in its entirety in "The Psychology of Astronauts" (46–48), and another sequence, in which Mailer first hit on the idea of the unconscious as a kind of memory bank, appears in revised form in the second segment of "The Psychology of Machines" (156–57).

Although obviously written in haste, and at a time when Mailer had not yet found his theme, these pages read well, and the numerous penciled insertions in his hand suggest that he had gone over the original typescript at least twice.[4] Except for a wholly mechanical transitional appearance or two, the Aquarius persona plays a muted role in this chapter, Mailer having crossed out every personal reference to himself and to his dream life. In effect, all that remains of Aquarius is the psychological profile of a compositional self, an abstract mental apparatus that in "The Psychology of Machines" would become the conscious Novelist and the subconscious Navigator. In the published text, as in "Alpha and Bravo," Mailer depicts the Novelist as that part of the psyche which is forever drawing and redrawing the charts on which the Navigator will base a "vast social novel" (*Fire* 157). So too does he speak of "that great novel in the map rooms of the self" in a 1967 article on film, parts of which may be regarded as a rough draft for the more extensive psychology Mailer creates in *Fire* (*Existential Errands* 113–14). But nowhere in the published text or in any prior work do the outlines of such a novel take shape quite so explicitly as in the "Alpha and Bravo" manuscript, and in few chapters of *Fire* is the loss

4. Mailer's first draft of *Fire,* like the initial drafts of all his books after *The Deer Park,* was written in longhand, in pencil. The "Alpha and Bravo" typescript is split up into four segments labeled 10 through 13. Whereas portions of segments 11 and 13 survive in the published text and to a certain extent bridge the gap between "The Psychology of Astronauts" and "The Psychology of Machines," the precise location of "Alpha and Bravo" in the original draft awaits further textual study.

of ego so nearly complete or the narrative consciousness rendered so detached and impersonal.

In this way, contrary to the stereotypical view we have of him, Mailer deemphasizes art as an expression of personality and admits, at least implicitly, that one cannot challenge the repressions and sterility of the technological age merely by imposing one's ego on the world. And yet, as always, Mailer approaches such total representations by way of the self, seeking to embody the nation's many levels and contradictions within the individual psychic apparatus:

> Left and Right. If in Europe they were born as in separate millennia, as profoundly opposed philosophical ideas of social balance and justice, now in America they symbolize our schism. On the left are hippies, weathermen, SDS Communists, Trotskyites, liberals, social democrats, trade unionists, democrats, blacks, militants, consumer groups, TV commentators, ad agencies, student peace groups, academics, half or more of the working class, and an enormous swatch of the urban middle class. . . . On the right is the marine corps, the pentagon, the aerospace industries, professional athletes, corporate executives, bigots, squires, country gentry, philosophical conservatives, radical reactionaries, minutemen, farmers, veteran organizers, astronauts, rednecks, small-town high-school principals, state troopers, city police, the majority of dioceses in the church, the other half of the working class, a near unanimity in the small-town middle class. . . . What is immediately apparent is the near indigestible variety within the left and within the right. Such ingredients in a human being would produce a personality doubtless hysterical, vomitous, disoriented, psychopathic, psychotic, wealthy, manic depressive, feverish, yet with it all productive, powerful, and possessed of extraordinary funds of unforeseen energy. (58)

Mailer acknowledges that "a social scientist who draws any relations promiscuously between the psyche of a man and a nation is no poet" (57), but the psychological model is the only way of holding together so many conflicting populations and interests—hardly a country at all. More likely still, the "unforeseen energy" of the psychopathic personality would force the nation apart. At the close of *The Armies of the Night*, for example, Mailer foresees a revolution growing out of the extreme psychological differences that had separated the warring "armies" on the steps of the Pentagon during the antiwar demonstration of October 21, 1967: "Whole crisis of Christianity in America that the military heroes were on one side, and the unnamed saints on the other!" (316). From the perspective of time, however, the crisis of Christianity looks more like a crisis of Mailer's own desire to narrate cold war history. For he continues, in *Armies,* to internalize cold war oppositions —left and right, totalitarian and liberal—that could never contain the nation's "indigestible variety" of voices, populations, cultural accents.

Throughout the book the Mailer hero, as a "left conservative," tries to probe the contradictions "at war" in his own personality in order to mediate between the embattled halves of America. The effort can be compelling, as when he considers the number of home, church, and media implantations that must be overcome before the most independent of Americans—not excluding himself—might allow themselves to cross a military line, or it can be altogether despairing, as when, riding a school bus down a prosperous shopping street in Virginia on the way to sentencing soon after his arrest, Mailer listens to the student activists with him yelling out slogans at random uncomprehending teenagers. Recalling himself to "that long dark night of the soul when it is always three o'clock in the morning," Mailer attributes both his present and F. Scott Fitzgerald's past depressions to the evident fact that, despite all literary efforts, "the two halves of America were not coming together, and when they failed to touch, all of history might be lost in the divide. Yes, there was a dark night if you had the illusion you could do something about it, and the conviction that not enough had been done. Or was it simply impossible—had the two worlds of America drifted irretrievably apart?" (177). For Mailer such sad reminders of American demographics translate all too readily into systematic psychological terms. As he would note with ever greater frequency during the waning years of the sixties: "We live in an American society which can remind you of nothing so much as two lobes of a brain, two hemispheres of communication themselves intact but surgically severed from one another" (*Existential Errands* 121).

By the time he wrote "Alpha and Bravo," Mailer seems to have realized that the country's supposed drifting apart was more the result of an overly simple model for the American psyche than a reflection of any actual breakdown in communication among individuals. In *Armies* Mailer's large formulations, binary groupings, and crowded lists are never so convincing as are his own encounters, both private and public, with participants in the event: the liberal intellectuals and writers who organized the march, literary friends and strangers on the way to the Pentagon who recognize him as the celebrity "Norman Mailer," and the unlikely collection of marshals, hippies, American Nazis, and professional revolutionaries he comes to know on the bus and in prison. Gradually and cumulatively, without his ever having to insist on his own totalizing systems, these meetings create for the reader a full sense of the many contentions and partial social reconciliations within the larger confrontation in Washington.

At the start of *Fire,* Mailer attempts a similar mediation among the many separate personalities who come together to witness the launch of *Apollo 11*. But here the very magnitude of the event and its near certain success, no matter how well or how poorly the spectators understand the technological details involved, make such personalities seem irrelevant. Except for a common vague awareness among the journalists, spectators, and corporate executives that their own achievements appear shoddy in comparison with the undisputed accomplishment of *Apollo 11*, it cannot be said that the social, economic, and racial strata come together compellingly in Mailer's imagination, any more than Armstrong sitting in the commander's seat of the lunar module can be said consciously to comprehend the "whole congeries of Twentieth Century concepts and forces which have come to focus that this effort may fly to the moon" (*Fire* 182). From so wide a range of classes, interests, and technical expertise does NASA draw its power that no single consciousness could possibly contain it all within itself—not and remain sane.[5]

Yet the astronauts, the sanest of Americans, somehow *can* contain such contradictions, more intensely even than psychopathic personalities. For these men must live

> with no ordinary opposites in their mind and brain. On the one hand to dwell in the very center of technological reality (which is to say that world where every question must have answers and procedures, or technique cannot itself progress) yet to inhabit—if only in one's dreams—that other world where death, metaphysics, and the unanswerable questions of eternity must reside, was to suggest natures so divided that they could have been the most miserable and unbalanced of men if they did not contain in their huge contradictions some of the profound and accelerating opposites of the century itself. (*Fire* 47)

5. Mailer has better success in his fiction, where a character's sanity may come to depend on "the ability to hold the maximum of impossible combinations in one's mind" (*American Dream* 158).

Unlike the open converse that occurs within the psychopath, the psychic communication Mailer posits within astronauts is not conscious. What makes them so exemplary is that their deepest motivation, like the dominant tendency of the age, remains unconscious, unknown to them.

The thesis Mailer originates here eventually appears only slightly revised in the final text of "The Psychology of Astronauts": that by exploring the unconscious life of the astronaut one might eventually come to comprehend the age's unknown "metaphysical direction" (*Fire* 48). What in the published version is left unstated, however, and what by now Mailer perhaps only suspects might be ineffable, or even nonexistent, is the hidden form of that internal communication by which the astronauts might contain within themselves "some of the profound and accelerating opposites of the century." We may also suspect such grandiose rhetoric of excluding more possibilities than it opens. It would seem to eliminate, for example, the possibility that Henry Adams perceives in his own attempt to postulate "a dynamic theory of history": "The Universe that had formed him took shape in his mind as a reflection of his own unity, containing all forces except himself" (475). The sublime equation joining and separating the self and these prior self-creative forces eludes Mailer, for however much he may limit his own role within the Alpha and Bravo system, he has not yet been willing, with Adams and such of his literary heirs as Pynchon, McElroy, and DeLillo, to allow a system of technological forces to take shape in his mind without first imposing his own imaginative structures on them.

It is not surprising, then, that the argument of "Alpha and Bravo" begins to falter at just this point: "To push further along these lines," Mailer writes immediately after introducing the idea of the unconscious, "is to recognize that talk of Alpha and Bravo will no longer suffice." The psychic apparatus, to be true to his model, must possess not one but two unconsciouses, Alpha id and id Bravo; and the health of the communication between the halves depends on the ability of Alpha id to speak not only to Alpha ego but across to "the ego of Bravo, just as the id of Bravo must in its turn be able to reach the consciousness of Alpha" ("Alpha and Bravo" 72). There follows an elaborate scenario of psychic division, conflict, and conciliation promising to rival Freud's tripartite division of the psyche into the ego, the id, and the superego:

> If love, health, and art [are] in large degree dependent upon the unconscious life of Alpha or Bravo being able to speak across to the ego

> of the other, what is one to make of that more arcane communication from depth across to depth, from Alpha id to id Bravo? Can it be as rare as the passage between planet and satellite? But in fact, suppose that on this thought we have come into sight at last of the lady with the silver mirror who dwells between passion and death. Can she be the true incarnation of the dream? Have we arrived at the harbors of the moon? (73)

The phrase "the harbors of the moon," the title image of the opening chapter of *An American Dream,* suggests again that what Mailer has been seeking all along is a fictive embodiment of the Alpha and Bravo configuration. The echo also indicates how far Mailer has come from the reality of the moon shot. At this stage in the composition the dream, for all its theoretical complication, has certainly fallen short of the overarching metaphor projected early in *Fire.* The romantic's desire for unconscious wholeness reclaimed and made complete in the creative imagination remains apart from his subject, and the Alpha and Bravo configuration comes to express a purely invented reconciliation where none exists, eventuating in a complicated set of metaphors that could never be taken for reality.

Mailer still falls back on prior imaginative connections when he could be building a real system, for there *were* other connections for him to build on, although he does not interest himself in the research of earlier psychic explorers which would have allowed him to see the connections. The dream, for example, has historically been likened to a machine in ways that are not incompatible with the Alpha and Bravo conceit. If we think of the machine, as Freud may have thought of it late in the nineteenth century, as "a system of opposed parts so arranged as to transform raw energy into 'work' " (Leed 42), then Mailer's dream is, like Freud's, a psychic machine, a structure of entities in conflict, arranged to transform libidinal energies into sources of value—"love, health, and art" ("Alpha and Bravo" 73). Of course, Mailer's dream metaphor differs from both the Freudian dreamwork and nineteenth-century conceptions of the machine in being a scenario or theater of simulation where existential possibilities are enacted. There is no mechanical certainty about how the various parts of the psyche interact, none of the technologist's limiting of communication to a measurable exchange of information. Mailer's unconscious is not the unconscious of Freud but the romantic unconscious "of imaginative creation . . . the locus of the divinities of night" (Lacan 24). And Mailer is not about to admit a technological or psychoanalytic definition that might in any way reduce the mystery of creativity, the ineffable psychic core around which his fiction is built.

This opposition between a living and mysterious psyche and more rational modes of structuring the world is of course solidly in the tradition of the romantic rebellion against the sciences. Yet I cannot help feeling that here Mailer loses more than he gains from the tradition. In "Alpha and Bravo" and the published chapters on psychology, Mailer seems almost willfully ignorant of work being done by practicing scientists and psychotherapists that might have deepened and advanced his own style of existential psychoanalysis—Julian Jaynes's study of bicameral brain economies, for example, Gregory Bateson's cybernetic models of social and mechanical communication, or R. D. Laing's passionate researches into schizophrenia and other psychic disturbances. Metaphor, for these scientific writers as for Mailer, does not simply describe consciousness. Rather, in Jaynes's formulation, it "generates consciousness" out of past experience by "constantly and selectively operating on such unknowns as future actions, decisions, and partly remembered pasts, on what we are and yet may be" (56, 59). This conception of metaphor is very consistent with Mailer's conception of the dream, not as Freudian wish fulfillment but as a mechanism of comparison between the past and the future that might prepare one for unexpected encounters, relationships, and heroic actions. Likewise, Jaynes's conception of the two hemispheres of the brain acting like "two independent persons, which in the bicameral period were . . . the individual and his god," could only have strengthened Mailer's psychology by introducing actual and intimate gods whose function was "chiefly the guiding and planning of action in novel situations" (117).

No less striking are the similarities between Mailer's social narratives and Bateson's view of the self and its relations with others. Especially relevant are his comments on the nature of control, by which Bateson means "the whole related complex suggested by such words as manipulation, spontaneity, free will, and technique" (267). When most successful Mailer's dream narrative enacts what Bateson would term a cybernetic "feedback loop," which, through "a process of *trial and error* and a mechanism of *comparison*," enables both mechanical systems and living organisms to adapt to changes in the outside world rather than simply controlling the world (274). As we shall see in the next chapter, Steven Rojack in *An American Dream* is made to deal with changes in the social environment in just such a cybernetic fashion: he improvises his reality as he must, inventing and rejecting future scenarios with equal ease, "as in a movie" or, more often, in a surreal reverie (51). Through endless repetitions and metaphorical accumulations these imaginary projections take on the vividness of reality; they are the trial-and-error processes and comparative mechanisms that give him a tenuous and adaptive sense of himself. The danger arises, however, when Rojack (like Mailer in *Fire*) opposes his psychic reality too strenuously to a power he perceives as all-consuming. Rather than regulating his relation to the social world in a give-and-take manner, so oppositional a mechanism is apt to oscillate out of all control, leaving him in the end dashing between apocalyptic extremes.

Mailer, then, may have already achieved the kind of narrative he failed to embody in "The Psychology of Machines," an imaginative exploration into levels of "reality subtly beneath realty" forced to the surface by the pressure of a rich, hyperbolic language (*Fire* 160). Through repeated dreamlike simulations he may be said to have created in his main protagonists an approach to psychic wholeness consistent with Bateson's "Steps to an Ecology of Mind." The holism of the fiction, however, is not one that can be achieved apart from the technological society itself. Not only do the novels represent the nature of social and technological control; they also reveal the limitations of any attempt, literary or otherwise, to oppose such control directly.

2

"Alpha, Omega" and the Sublime Object of Technology

That Mailer's 1991 novel *Harlot's Ghost* should end on page 1282 with the words "TO BE CONTINUED" is perhaps the most fitting fulfillment of his long-projected novel of America. From the time of *Advertisements for Myself* (1957), when Mailer first announced his conviction that his "present and future work" would have "the deepest influence of any work being done by an American novelist in these years" (17), a source of stimulation for his best writing has always been the *next* project, which would usually be abandoned the moment he completed the work at hand. He wrote perhaps his finest short novel, *Why Are We in Vietnam?*, under the impression that it would be merely a prologue for the promised big book, and he thought that *Of a Fire on the Moon* would likewise develop into a massive psychological narrative, the ghost of which can be read in the discarded "Alpha and Bravo" chapter. In retrospect the very real accomplishment of *Advertisements* must be seen to depend largely on its many unfulfilled promises; without them Mailer might never have done as much as he has.

In *Harlot's Ghost* the old restlessness is still in evidence, although much attenuated by an expansive, rather formless prose. Mailer's preoccupation with Alpha and Bravo (here they are called Alpha and Omega) is given over wholly to one character, Hadley Kittredge Montague, whose marginal life in the Central Intelligence Agency leaves her wishing "in some most determined part of [herself] to become [her] own intelligence center" (1019). Something like becoming one's own intelligence center has of course been Mailer's ambition from the start, and in *Ghost* Mailer continues to bring to the task an impressive knowledge of corporate and international life. In the course of a long career, however, he is no closer than Kittredge is to the secret mind of America: "In truth," she writes, "I need to know much more about everything if I am ever to be able to write about Alpha and Omega in many walks of life. I get out, of course, and I do meet people, but I know so little of how the real gears work in that hard fearful real world out there" (1075).

In Kittredge's dualistic psychology Alpha and Omega represent conflicting aspects of the human psyche, like separate personalities living in a single person, whose peculiar tensions are thought to embody contradictions and differential relationships within society at large —contradictions that can be only dimly perceived, if not resolved, through an individual's confrontations with various social and psychological Others. Her psychology has mostly to do with "metaphor," a process in Mailer whereby spiritual forces are sublimated into everyday topical terms familiar to the reader of American newspapers and the viewer of multichannel media. From the start, his narrative is cast as a fiction of oppositions, a metaphorical discourse grounded in the impossible desire to bring his private literary struggles—with his precursors and with his own earlier accomplishment—into the larger arena of contemporary power, even into the very media he would attack. The attempt, of course, like all sublime attempts at literary transcendence, is doomed to failure, as his characterization of Kittredge implies. For those in power the "metaphor" of her psychology is taken to mean only that Alpha and Omega must be "a whole-cloth projection" of her own "latent schizophrenia," and her theories, not surprisingly, are soon overshadowed within the CIA bureaucracy by more pragmatic techniques of psychological manipulation and control (560). Defeated by such literal-mindedness, she fears that Alpha and Omega may have become "last year's intellectual fashion" (625).

Is this Mailer's fear as well? Kittredge's self-criticism repeats a charge that has often been made against Mailer. It is just "the lack of a convincing vision of how things actually connect in the world" that Gerald Graff once ascribed to him in the course of a powerful critique of contemporary writing in general (*Literature against Itself* 219). Graff discerns a tendency, "already present in classic modern fiction" and carried to an extreme in "postmodern" fiction by Mailer, Thomas Pynchon, and William S. Burroughs, to improvise a "private typology" in the face of antithetical communicative forms, a tendency, Graff perceives, that only increases the writer's alienation and that accounts for the inability of much contemporary fiction "to retain any moorings in social reality" (*Literature against Itself* 209). Graff needs only to point to the anticlimax of Steven Rojack's confrontation with the industrialist Barney Kelly, the admitted "focus of evil" (Mailer, *Pontifications* 80) and politically the most powerful character in *An American Dream*, to make us feel the arbitrariness of many of Mailer's psychic connections. Rojack's conversation with Kelly, culminating with the famous walk around a parapet high in the Waldorf towers, does seem a wholly private act of defiance, "its significance remaining unassimilable to any publicly shareable system" (*Literature against Itself* 219). Yet it is possible that the connections in *An American Dream* point to a more complex conception of power than any that Kelly as a single representative of public institutions might embody, however near he may be to what he calls the "magic at the top" (246).

Nowhere is Graff willing to consider technology, for example, as a way of representing networks of contemporary power that our minds and imaginations have great difficulty grasping verbally. What Graff dismisses in Mailer, Pynchon, and perhaps rightly in Burroughs as "a diffuse, semi-mystical enthusiasm for the vibrating electrokinetic dynamisms of technological society itself" may be, without Graff's deliberately obfuscating rhetoric, precisely what is needed to achieve a contemporary "vision of how things actually connect in the world" (*Literature against Itself* 217, 219). Graff, among many others, objects to the paranoia that typifies a protagonist such as Rojack, whose sanity may come to depend on "the ability to hold the maximum of impossible combinations in one's mind" (*Dream* 158). This distinctly *literary* paranoia, however, in Mailer no less than in Pynchon, is not a clinical condition or even a psychological state so much as it is a conceptual commitment to systems of connection—the "like knowledge" that is the word's literal meaning. The high magic and underworld paranoia experienced by Mailer's protagonists become metaphorically linked with real struggles. Both generate psychic undercurrents that give form to the American epiphenomena—the endlessly fascinating experiential and journalistic details—that would otherwise crowd the pages of all Mailer's work and bury it in mere topicality.

To be sure, the private imaginative history that would embody so much American matter is not ultimately "shareable" in Graff's sense of being amenable to public discourse or debate, for the power of such a narrative can be presented only insofar as it *exceeds* the phenomena that the imagination takes as its material. Neither will the imagination's terms—its gods and devils, its concern with natural and demonological forms that transcend the human—ever be brought into mediated discourse (however much the media themselves depend on nonrational stimulations to produce a rapt audience). "A humanistic sublime," let alone a politically correct one, "is an oxymoron," writes Thomas Weiskel (3). It is as much a precondition of the sublime as of a political orientation or temperamental idiosyncracy that causes Mailer to oppose himself to any humanistic or, worse yet, "liberal" ideology. Without an almost desperate antagonism and the promise of some dialectical movement toward reconciliation in the mind of his hero, Mailer's postmodern version of the "dynamical," egotistical sublime could not exist.

Reading *An American Dream* and *Why Are We in Vietnam?* in retrospect, with a knowledge of Mailer's speculations on the dream, "Alpha and Omega," and the "psychology of machines" (from the second part of *Fire*), we can begin to appreciate the full psychological complexity of Mailer's fiction and to evaluate his novels within the context of contemporary explorations into the self, the sublime, and the technological subject. Central to his entire work has been the desire to intensify and reconcile opposites, a dialectical movement toward a unified consciousness that is not quite what Richard Poirier has in mind when he visualizes Mailer "at war" with himself (or, more precisely, with the sources of American social and political power that tend to repress minority elements within the self). Although the Mailer hero is certainly combative at times, and war is a recurrent theme especially appropriate and accessible to Mailer's audience in his own generation, I believe that war is essentially a figure for a literary struggle that is doomed to failure, signifying precisely the imagination's inability to comprehend or represent the *other* media that are believed to dominate contemporary American life. The "dynamical sublime," as Weiskel notes, "is always cloaked in metaphors of aggression" because it seeks to embody power, which is not in itself literary, in discourse (5). One finds in Mailer a rare combination of the rhetorical (or literary) and the technological sublime, in which language is *set against* a nonliterary construction, a symbol of the mind's relation to an all-powerful dominating order.

The agitated resistance in Mailer to all manner of "totalitarian" social and political forms is, I believe, the same aggression that fuels his literary struggle, his need to outdo not only his literary contemporaries, and not only Hemingway and the great naturalistic writers who helped form his early ambitions, but also his own work as he moved away from the social realism and satire that he had assimilated from Dreiser and Dos Passos in *The Naked and the Dead,* from Malraux in *Barbary Shore,* and from Nathanael West in *The Deer Park.* Poirier rightly perceives an internal division in Mailer—a "minority within"—that must constantly resist his own as well as these antecedent literary forms and metaphors (see *Mailer,* chap. 2). But war and self-division do not by themselves explain his need to find a new style for each new book, or the compulsion to write each book while on the way to some other book that is projected but never completed. Less conclusive than any war, this detour followed by romantic epiphany is a characteristic experience of the sublime, whose deeper movement is syncretic or relational, an adventure of consciousness that brings together opposed social classes, personalities, and conflicting modes of thinking and experiencing.

Such struggles must be taken with a certain irony today, especially now that both the actual war in which Mailer fought and the existentialist quest to write the great social and psychological novel of America have themselves become historical. But it is important to note that, outside of public pronouncements in his capacity as portly president of PEN and general literary statesman, such an irony is mostly lacking in the novels themselves. Insofar as this absence causes him at times to take himself too seriously, the lack of irony is surely a flaw, but it is a flaw that might explain what real staying power Mailer does possess. After the particular occasions for his embattled existentialist metaphors have passed, he continues to write as if literature *mattered.* He refuses the easy acceptance of an irony that not only kills the sublime by reducing it to bathos but also too easily masks a bland acceptance of the consensus culture. Irony is certainly anathema to a literary project "which concerns power and sets man and nature in desperate opposition," as in Weiskel's definition of the "dynamical sublime" (5). This project also concerns "desire"—not only the desire to stand above one's literary competition but the desire to join with another, with anything outside the self that might be called an *object.* It may well be a condition of Mailer's continuing dynamism that, like all romantic desire, the unification of subject and object will never be fulfilled. But Mailer would be the last writer to admit that such fulfillment, the merging of mind and world into an identity beyond all alienation, is perhaps as unlikely as it is *un*desirable.

Mailer's Melodrama of Beset Manhood: *An American Dream*

"There is no such thing as the social production of reality on the one hand, and a desiring production that is mere fantasy on the other," write Gilles Deleuze and Felix Guattari in the "Desiring-Machines" chapter of *Anti-Oedipus.* That is, there is no basis for regarding one's psychic reality, the mental connections produced by language and desire, as any less real than technology's linking of disparate data, or than any of the more official linkages within the "social machine" (28). Indeed, such desire inhabits the technological world. Fantasy and dream are not means of escaping the real; they are on its side, and essential to its production.

An American Dream might be read as an American precursor and approximation to the "social field . . . immediately invested by desire" analyzed by Deleuze and Guattari (29), a field of dreamlike repetitions and differential relationships in which the self schizophrenically lives out the contradictions of contemporary reality. Not *simply* the result of a private subjectivity, the fantastic connections and acts of stylistic daring in the novel proliferate in the felt absence of any immediately locatable center of power in the world; yet the connections need be no less real or "convincing" for being the product of a character's desire.

Where the modern writer runs into difficulties, however, is in supposing that the reality of desire can be used *against* an impersonal, technological reality. In many ways Mailer's rhetoric of desire is consistent with the liberatory rhetoric of sixties activism. But the commercialized, increasingly technologized world of succeeding decades has proven extremely efficient at turning the pursuit of desire to profit. " 'Take your desires for reality!' can be understood as the ultimate slogan of power," writes Jean Baudrillard in sobering counterpoint to Deleuze (*Selected Writings* 179). What once appeared liberating has by now grown oppressive, and the Mailerian-Deleuzean pursuit of desire may actually have helped to turn the contemporary world into a mirror, a simulacrum of Self whose apparent freedom from constraints generates a narrative whose only imaginable end is in death.

Without doubt, a deathward orientation accounts for the bizarre atmosphere of *An American Dream,* in which Manhattan is as alive with strange forms, spirits, and demons as D. H. Lawrence's Etruria or, later, the Egypt of Mailer's own *Ancient Evenings.* The incantatory "voodoo narrative" of *Dream,* which T. H. Adamowski finds "as calculated as the anti-intellectualism of *Etruscan Places* or *Fantasia of the Unconscious*," is intended, no less than the redemptive Reichean orgasm Mailer so often invokes, to break down our ways of looking at reality, particularly as the real is defined by the rationalized professional classes depicted in the novel, or by the liberal academic and professional intelligentsia that Mailer vilifies in *The Armies of the Night*. These academic and professional groups depress Mailer because they are all of a piece with "technology," more or less willing "servants of that social machine of the future in which all irrational human conflict would be resolved" (*Armies* 25). In *An American Dream* such conflict is purposely intensified and allied to the ghosts and demons that Mailer's opponents in the technocratic liberal camp can only treat ironically. Deliberately melodramatic, as "the melodramatic imagination" is defined by Peter Brooks, the novel presents a world where spiritual forces and imperatives are operative, in which Rojack is made to live out social, ethical, and psychic dramas that continually bring to the surface forces that have been all but banished from the secular liberal mind.

Less generously, we might also identify Rojack's progress in feminist terms, as a "melodrama of beset manhood" such as Nina Baym finds dominating American literature. Baym sees the typical American male protagonist as forever in flight from "the destructive pressures of an overcivilized, artificial society identified with women" (Graff in *Professing Literature* 222, summarizing Baym). And it is true that women in *An American Dream* are perceived by Rojack as the real source of institutional power in America. From the moment in the opening chapter when Rojack murders his wife, Deborah, for having "occupied" or "owned" his "center," he is continually made to resist psychic and physical possession by various sexual, racial, and social Others, all the while absorbing something of the good and evil in each (*Dream* 27, 176). His sexual domination of Deborah's maid, Ruta, who during the act is said to want "no more than to be a part of [his] will" (45), can be read all too easily as an attempt to overcome male powerlessness by subordinating female power, the woman being presented not as an individual but as a collection of psychic forces, mysterious procreative powers, national and class attributes. (Absurdly, Rojack desires anal intercourse with the German Ruta for the "high private pleasure in plugging a Nazi" [44]).

Even Rojack's new lover, Cherry, who evokes every American blonde (or artificial blonde) from Marilyn Monroe to the girl next door, is presented sentimentally as a "nest of separate personalities" whose sum equals America (97).[1] Cherry's attraction is to be, like Susan Alexander in Orson Welles's film *Citizen Kane*, a "cross-section of the American public," whose seduction might confirm the entrepreneurial male's ability to master the raw material of American culture. The passage in which Rojack removes Cherry's diaphragm during sex without regard for her desires would appear to be one more instance, after Deborah and Ruta, of willful mastery, almost a parody of the American male writer's stereotyped desire to bring women back to "the imperatives of their gender, which require marriage, childbearing, domesticity" (Baym 75). To be sure, Mailer depends on sexual stereotypes to outrage liberal readers: the extreme violence of Rojack's sexual encounters, like the predilection toward violence and the tiresome macho posturing in Mailer's public life during the sixties and early seventies, seems to have been designed to bring forth equally stereotyped interpretations of *Dream*'s war between the sexes. Judith Fetterley, for example, finds the novel's sexual ideology illustrative of "the phallic critics' mythology of an embattled male, suffused with fear, fighting off the malign influence of witchy, bitchy women" (136). But if we are not to measure Mailer's sexual politics by the very conceptions of power he means to resist, it is necessary to go beyond the stereotypes.

To begin even a negative critique of Mailer's sexual politics, which I am here reading as a sublimation of a dominant technoliberal ideology into a metaphor of unconstrained male desire, one must first recognize the proper object of his anxieties. Fetterley connects Rojack's distaste for birth control and his insistence on "the crucial importance of the vaginal orgasm" with his fears about his identity, including "the fear that he is not necessary for women's sexual pleasure" (141). Yet Rojack during the act is hardly concerned with pleasure, performance, or his isolated identity; pursuit of the first can produce only repetitive dullness for him, and he has also become bored by the sundry performative tasks that would make him a worthy partner in the eyes of Deborah, the heir to Kelly, and that he might be taking on with Cherry, Kelly's onetime mistress. Rather, the diaphragm's removal is necessary for both lovers to lose themselves in the act, to forget their social roles and dissolve their independent wills. The diaphragm, "that corporate rubbery obstruction beam," is the most direct representation in the novel of technology's incursion into the human body, which as

1. Cherry's original, Mailer's third wife, Beverley Bentley, would be treated similarly in *Armies* (190–92).

a consequence has become separated from both psychic and sexual experience (127). Mailer objects to modern technologies of birth control, here and later in *The Prisoner of Sex*, but not because they limit male autonomy by giving women greater control over their own bodies. Autonomy achieved through rational control and power conceived as an expression of will are precisely what this passage would move beyond.

Mailer's attack, then, is not against women but against an excessively *masculine* culture shot through with technology and a social production of reality that has little to do with desire and everything to do with control. It is Mailer's belief that technologies—especially sexual technologies—distance us from our bodies, from necessary anxieties about both personal mortality and procreative sexuality, and so create a climate suitable to the kind of organized social control that makes units of people.[2] His defense against critics who would dismiss not only this passage in *Dream* but comparable passages in the work of Henry Miller and D. H. Lawrence is that such criticism, by insisting on a feminine self no less willful than the competitive male consciousness it engages, ends by contributing to a climate of sterile rationalism, fractured egotism, and technological manipulation.

Mailer himself, of course, is not immune to charges of egotism, and I am not suggesting that his attempt to get beyond the enclosed, protectively rational self is wholly successful. His antifeminist critique, though once powerful, has grown dated next to more subtle feminist engagements with technology of the eighties and nineties. Linda Brigham's critique of the French theorist Paul Virilio, for example, is much more perceptive and to the point than any feminist attack yet leveled against Mailer. Like Mailer, Virilio sees feminism as a product of the technologization and resulting destructive speed and "motion" of everyday life. In Virilio's account, writes Brigham,

2. In *Prisoner*, for example, while discussing the Kinsey Report on the orgasm, Mailer asks: "What desire had technology to calibrate this being-within-a-being when the human was the unit, and the groupings of unit were blocks of social use?" (88). Mailer's objection, as in this passage from Muriel Cerf's novel *Street Girl*, is based on a subtle distaste for the desensitizing of *language*: "Although nobody even mentioned sex to me before, all of a sudden they started stuffing me full of all this talk, both vague and precise, making precise what should be left vague and vice versa, small masterpieces of turdified obscenity that would have made me frigid for the rest of my life if I hadn't let it go in one ear and out the other. . . . Sex, did you say? Excuse me, you're talking about the uterus, ovaries, the penis and lubrication, don't know what you're getting at, save those words for Molière's quacks, they smelled like church and wet-nurses' milk." (174–76)

> the feminist *movement* is a movement toward masculinity and the annihilation of more natural gender distance. This line of argument is now a quite familiar antifeminist one: in deeply questioning the trajectory of technology, [Virilio] condemns the alliance of women with it as doubly unnatural. In the light of alternatives such as that provided by Donna Haraway in "A Cyborg Manifesto," this may well represent only a failure of nerve, a fear of the unknown, and a deeply reactionary masculinism. (10)

Indeed, it is just such a fear of the unknown, or rather an insistence that the natural world remain unknown and deeply identified with a feminine "mystery," that ultimately limits Mailer's engagement with technology, in *Dream* and in later novels. As a survey of his representation both of women and of nature throughout these novels shows, with the exception of *Why Are We in Vietnam?* Mailer has not met Haraway's contemporary challenge "to live on the boundaries, to write without the founding myth of original wholeness, with its inescapable apocalypse of final return to a deathly oneness that Man has imagined to be the innocent and all-powerful Mother" ("Manifesto" 176). By insisting that the feminine remain aligned with nature, and by setting that nature in an oppositional rather than integral relation with technology, Mailer is left un-ironically generating, in Haraway's words, "antagonistic dualisms without end (or until the world ends)" (180). Haraway's description is perfectly consistent with the description given earlier of the dynamical sublime, which characteristically features an endless obsessional proliferation of dualistic oppositions, thus producing a complicated set of metaphors that can resolve the oppositions only in some apocalyptic, supersensual destiny. Those who, like Mailer, work in this version of the sublime thus tend to have an especially difficult time imagining any possible ending for a narrative other than death. For, as Baudrillard argues, death is the sole remaining absolute that can stop the endless play of signification in a nonreferential culture of media simulations and unconstrained consumer desire.

Consider the one moment in *Dream* (besides the scene in which Rojack removes Cherry's diaphragm) when technology is foregrounded. Soon after Rojack has been escorted by the police detective Roberts through a hostile gathering of the press, he is led to the morgue to identify Deborah. In the corner of the room he notices a television set "turned down low and . . . out of synchronization, for the picture was flaring bright, then dark, then flaring up again, and I had the insane clarity to recognize that it was speaking to the neon tubes and they were answering back" (76). Deborah has been, and will continue to be, a presence for him in death no less than in life, but this ghostly communicant is now replaced by a technological dialogue wholly unsynchronous, not only with itself but with the imaginary dialogues and natural communions that Rojack has felt all this time wedding him psychically to the woman he has killed. For having "pushed Deborah into the morgue," and for this only, does Rojack feel guilty:

> Whatever Deborah would deserve, that morgue was not the place for her. I had a reverie of my own death then, and my soul (some time in the future) was trying to lift and loose itself of the body which had died. It was a long process, as if a membrane trapped in mud were seeking to catch a breeze which would trip it free. In that morgue (for that was where I pictured my own death) the delicate filaments of my soul were also expiring in a paralysis of deodorant while hope withered in the dialogue between the neon tube and the television set. I felt guilty for the first time. It was a crime to have pushed Deborah into the morgue. (76–77)

This passage is remarkable as much for its intertextuality as for its content. Rojack's reverie of his own death, recalling as it does his vision of a death-in-life on the balcony at the start of the novel, also looks forward to the disembodied narrator early in *Fire* and to the prolegomenon of *Ancient Evenings*. A certain cluster of language and imagery gets repeated in each passage: "the delicate filaments" of Rojack's soul return in the later novel as a "double filament" expiring in the effort to free itself from Egyptian mud and clay and the "screaming scalding wastes" of Menenhetet II, the dying hero's, flesh (*Evenings* 4). Menenhetet at each recurrent seizure of pain laments "the lost dialogue that [in life] had never taken place between the bravest part of me and the rest" (4). Rojack will similarly retain the memory of his psychic dialogue with the moon, for which the present "dialogue between the neon tube and the television set" is a parodic substitution.

Against such disembodied electronic dialogues, Rojack's reverie and Menenhetet's struggle in death are ways of recovering sensual life and a psychic wholeness that Mailer feels has passed from the contemporary scene. The dialogue "in the burial chamber of the Pharaoh Khufu" allows Mailer to imagine a consciousness that is "dead yet more alive than before" and that ultimately will return to a body capable of more fully sensing a world of sight, taste and "moldering smells [in the] heavy marrow of the air" (*Evenings* 107). More freely than in *Dream*, Mailer in *Evenings* would have the individual consciousness, at once acknowledging and withstanding the consciousness of others, contain within itself and *participate in* the tensions and interrelations of an entire society through a metaphorical language that intensifies consciousness by multiplying connections.

Integral to all such formal innovation is the idea of "communication," by which Mailer means not a technological linkage involving "machines and electronic apparatus and services of distribution" but the life that comes "from meetings and confrontations, from opposites coming together" (*Cannibals* 281). Unlike the metaphorical and creative meeting of opposites within the mind of a narrator-hero, technological communications imply for Mailer "the injection of information into a passive being" (*Cannibals* 281). Where the one, psychological communication becomes (ideally) the basis for imagining an integrated society, the other, technological relocation of information tends most often to "outrace the psyche" (as Mailer, at the very start of his career, wrote of the increasing mechanization of postwar American society [*Naked* 391]. Even in Egypt of the twelfth century B.C., where technological activity is limited to the construction of roads, the invention of a sluiceway for plumbing, and the quarrying of stones, the dynasty is weighed down by "dull fact" and weakened by relativized and competing voices (*Evenings* 198). "We do not have the desire to build one Great Pyramid," complains the Pharaoh Ptah-nem-hotep. "We spend our lives on a hundred matters." The thoughts of the Egyptian population, which the Pharaoh must know if he is to possess his dynasty, produce only a din in his ear, the result in Mailer of all technological mediations that separate information from knowledge.

Much the same din is perceived by Harry Hubbard in *Harlot's Ghost* when he visits the great secret tunnel that CIA bureau chief William Harvey had built to monitor telephone conversations on the Soviet side of the Berlin wall. Hubbard, fresh from Yale University and his subsequent Company training, describes the "large windowless room with fluorescent lights overhead" (304) in which recording machines can be heard stopping and starting automatically:

> The sound of 150 Ampex tape recorders—Mr. Harvey provided the number—moving in forward or reverse, electronic beeps signaling the conclusions or commencements of telephone conversations produced

> an aggregate of sound that stirred me in the same uneasy fashion as some of the more advanced electronic music I had listened to at Yale.
>
> Was there one telephone dialogue between the East German police and/or the KGB and/or the Soviet military that was not being captured at this moment on one or another Ampex? Their humming and whirring, their acceleration and slowdown, were an abstract of the group mind of the enemy, and I thought the Communist spirit must look and sound like this awful room, this windowless portent of Cold War history. (*Ghost* 305)

This passage may be somewhat exceptional in Mailer's more recent work, not only for its technological content but for a perceptual intensity that is rarely permitted Hubbard as narrator. That Mailer in this late novel tones down his customary linguistic excess could well be a sign that he has mellowed with age; more important, it is also an indication of his continued stylistic integrity and precise sense of his own limitations as a writer. Mailer's strength has always been to adapt his style and even his enduring obsessions to the design-governing consciousness of a given narrator, and *Harlot's Ghost* may stand as the monumental literary survey of the cold war mentality in America. What seems to have reduced Mailer's aesthetic achievement, however, here and throughout the late work, is an inability to find a form for his fiction that approaches even the CIA's technological "abstract of the group mind of the enemy" (305). The subject of *Ghost* is of intrinsic interest, no less than was the story of convicted murderer Gary Gilmore in *The Executioner's Song* (1979), but the data Mailer unearths and the considerable American life he knows are somewhat laboriously handled in both novels. Even the haunting paragraphs of *Song*, each one a "self-contained dramatic unit" surrounded with "a generous aura of blank space" (Edmundson 440), often have the feel of having been worked up directly from notecard entries.[3] The narrative of *Ghost*, in spite of a complex framing device, is very linear and too drawn out for Hubbard's present perspective (in the "Omega" manuscript) to exert any formal tug on the past events he describes (in the much longer "Alpha" manuscript). Without the pervasive dualities of cold war politics that produced the content of *Ghost*, and without its apocalyptic prospects to give play to the writer's melodramatic language, Mailer

3. I treat *The Executioner's Song* in greater detail in chapter 7, in connection with the documentary realism of William Gaddis's *JR* and Don DeLillo's *Libra*.

is left to write an admirable though necessarily inconclusive history whose only possible ending, whether or not he finishes the project, is "TO BE CONTINUED" (1282).[4]

World without End: The Vietnam Novel

Nowhere in *Harlot's Ghost* does Mailer come near to either the linguistic virtuosity or the economy of form that he had already achieved in a much earlier novel, *Why Are We in Vietnam?* (1968). Not coincidentally, *Vietnam* is the only one of Mailer's novels that embraces the forms of technology directly, so as to erase the artificial distinctions between nature and technology, between the imagination and the reputedly mechanical processes of social communication and production. And yet, strangely, of all Mailer's novels *Vietnam* is the most like *Evenings* in its evocation of a vast natural anima—"the Magnetic-Electro fief of the dream"—which takes to itself the totality of human thought (*Vietnam* 170). For this reason, though it is certainly a lesser work—in scope if not execution—than either *Song*, *Evenings*, or *Ghost*, the Vietnam novel represents a more complete realization of Mailer's dream narrative.

There is little in Mailer that is likely to inspire critical consensus; and while evaluations of *Why Are We in Vietnam?* remain singularly contradictory, most readers agree in viewing the book, for better or worse, as a departure for Mailer.[5] One reason for this uniqueness is that Mailer wrote the book, as he says in a preface to a later edition, "under the mistaken belief I was writing not this kind of novel but another." The story about hunting bear in Alaska was initially meant to be not a full-length novel at all, but a prologue to a novel about a murderous group of "bikers, hippies and studs plus a girl or two living in the scrub

4. As the present volume goes to press, Mailer has completed a massive book about Lee Harvey Oswald and another on Picasso. He is said to have resumed work on the second volume of *Harlot's Ghost*.

5. Roger Ramsey notes with surprise the role played by technology: "From D.J.'s opening quotation of Edison, technology pervades the imagery; strangely, unlike Mailer, D.J. seems not to object" (418). Richard Poirier admires the "passionate energy" with which Mailer displays his mastery in this book, not only of the language of technology but of all "those expressive modes which threaten to obliterate his own expression" (*Mailer* 128). J. Michael Lennon, surveying the whole of Mailer's ouevre, finds the book "tangential to the principal curve of his development, which through the sixties continued to spiral inward." Lennon cites Mailer's admission: "I was full of energy when I was done, but the work was by the same token impersonal" (10).

thickets" outside Provincetown, Massachusetts, where Mailer had a summer home (*Pieces* 9). The early drafts of "Chap One," the opening chapter of *Vietnam*, and an abortive attempt by Mailer at the end of the book to move D.J. and Tex east from Dallas to the sands of Cape Cod, substantiate Mailer's claim that he wrote the book, almost by accident, while on the way to something else.

From experimental detour to romantic epiphany is a recovery typical of the sublime: "the obstacle in the way," as McElroy remarks, is often "the way" in postmodern fiction ("Midcourse Corrections" 14). But the *Vietnam* manuscript (stored in the warehouse on Sixty-second Street with the manuscript of *Fire* and the rest of Mailer's papers) suggests little of the long novel Mailer intended to write. The original opening paragraph (which reduces to the "Hip hole and hupmobile" paragraph of the published text) locates D.J., the narrator, in "Provincetown, where America began." At the age of twenty-two, four years after the published text's narrative present, D.J. presumably has been to Vietnam and is "now . . . in Provincetown," where "some of us live in the hills and make raids on the town—theft, robbery, mayhem, murder, raids on the very town where the pilgrims first came to port." A brief description of the way Cape Cod curls in on itself—which anticipates the northeastern geography at the start of Mailer's real murder mystery, *Tough Guys Don't Dance*—leads to the predictable obscene conclusion that if "Provincetown is the tip of the little tail of America, then what does that make Cape Cod Canal (alias Buzzards Bay) but America's own asshole—Hip hole and hupmobile."

Provincetown, however, was not all that obsessed Mailer. There was one passage cut from the manuscript draft that was more important than Provincetown to the book's emerging structure. The bulk of this passage—which in the published text reduces to one cryptic remark about "cunt and ass" (*Vietnam* 9)—is designed to complicate the narrative identity by hinting that D.J. may just have been subjected to an execution, and we are being told this story by an "expiring consciousness [that exists] here on the other end in death row." This is the deleted passage: "You are reading the words—if cunt is the sea and the earth the ass, then fire is time and the ocean is natural our future, for the land is our past, the air we breathe is present, present tense, you're full of fire, you are my expiring consciousness, am I a deadly man, no, no, no, dead" (manuscript 4).

Like Rojack before him and Menenhetet after, D.J. was thus conceived as a dying consciousness desperate to communicate a vision of existence from "the other end." Against a feminine, elemental organicism of earth, sea, fire, and air, we have male technology in the form of penicillin, electricity, and antibiotics ("penis-hill anti-bi-heartache" in the manuscript); and much as Rojack sees his soul "expiring in a paralysis of deodorant" in the city morgue, D.J. imagines his soul "expiring like that" under a technological assault on death, disease, and nonhuman nature (manuscript 5). Sense experience, in this deleted self-plagiarizing passage, remains opposed to technology, the very opposition that kept Mailer from reconciling the personal form of a psychological narrative with the technological society in *Dream*. Indeed, the only acceptable reconciliation would be a romantic merging of mind and world, and a loss of the physical self in the consoling arms of a feminized "nature" free of all technological incursions. The impossibility of such reconciliation, as we have seen, leaves the imagination to "generate," in Haraway's words, "antagonistic dualisms without end (or until the world ends)" ("Manifesto" 180).

Mailer had already identified most of *Vietnam*'s themes in the prefatory essays of *Cannibals and Christians*, although, as Tony Tanner immediately noticed, nowhere in this most exploratory of Mailer collections does he provide the new form that the book's antitechnological, antitotalitarian "Argument" demands ("In the Lion's Den" 55). At this still transitional stage between *Dream* and the composition of *Vietnam*, despite his early toying with electronic perception and an aesthetic of "distraction," Mailer was still stuck in a mostly linear narrative and a dependence on the traditional organic imagery of wholeness. As he entered more fully into D.J.'s consciousness, however, he would break up these sterile dichotomies to form nonhierarchical and continually shifting patterns. He was able, that is, to wrest a new aesthetic from the monotonies, broken moods, electronic and mechanical annoyances that for him characterize modern experience, an "art of the absurd" that in dealing "with categories and hierarchies of discontinuity and the style of their breaks . . . binds shattered nerves together by shattering them all over again with style, with wit, each explosion a guide to building a new nervous system" (*Cannibals* 247).

The one unifying metaphor in *Vietnam* that brings together both technological and natural fields of communication is a "new concept" of Mailer's invention, known variously as the "psychomagnetic field," "Magnetic-Electro fief of the dream," or, by abbreviation, M.E.F. (*Vietnam* 114, 170). Following McLuhan, for whom electronic communications promised to function as a global nervous system, Mailer takes the global village as a metaphor for a modern tribalism. D.J.'s language, moreover, is riddled with McLuhanisms, to the extent that McLuhan's favored dualisms—of hot and cool, explosive and implosive (chapters in the form of "Intro Beeps"), continuum and mosaic, medium and message—at times take precedence over Mailer's own. As much a totalizing image as any in McLuhan, the psychomagnetic field is yet another of Mailer's fictive embodiments of a national psyche, that bed of contradictions that was enabling America to pursue such happy images of a new tribalism, golden age, and global community while at the same time making war on specific populations in specific jungles and villages.

The novel itself, of course, is never so explicit about the war in Vietnam, though a year later in *The Armies of the Night* (1968) Mailer would risk an answer to the question posed by the novel's title: "We were in Vietnam because we had to be. Such was the imbalance of the nation that war was its balance" (209). The argument of *Armies* posits an irreconcilable conflict in the "average American" subscribing simultaneously to Christianity (whose center was "a mystery, a son of God" [210]) and the corporation (whose center was "a detestation of mystery, a worship of technology" [210]). The primary conflict in D.J., though similar, is altogether more personal and particular, and grows out of that final disillusionment with his father, who, after going off "on a free" with D.J. to find and kill a bear, would betray the experience on returning to the camp by claiming the fallen grizzly as his own:

> Between D.J. and Rusty it is all torn, all ties of properly sublimated parental-filial libido have been X-ed out man, die, love, die in a diode, cause love is dialectic, man, back and forth, hate and sweet, leer-love, spit-tickle, bite-lick, love is dialectic, and corporation is DC, direct current, diehard charge, no dialectic man, just one-way street, they don't call it Washington D.C. for nothing, eighteen, it's all torn, torn by the inexorable hunt logic of the Brooks Range when D.J. was sixteen. (126)

D.J.'s disaffection is total; but if it extends at all to the value system that Rusty, "the cream of corporation corporateness," represents (29), D.J. is nonetheless filled with the same brutal, one-way force Mailer perceived at the heart of corporate America. Fredric Jameson, reflecting on the seriousness of D.J.'s defiance, concludes that "it would be a mistake to think that D.J. rejects the value system of his father's existence (but it is a mistake of which D.J. himself is probably just as guilty as some of his readers): on the contrary his cult of fear and of courage is a way of being *more* faithful to those values than Rusty himself" ("Hunter" 192). Mailer's fiction, for Jameson, creates "some hypostasis of *competition* itself as a social and historical mode of being" (193), an antagonism of wills that we have seen Mailer fall back on whenever he cannot know the true power relations among individuals in society.[6]

The issue here goes beyond the success or failure of Mailer's aesthetic in the novel. Jameson perceives, but hastily dismisses, an "occult element" in *Vietnam* that functions as "a kind of *stylistic superstition*" or "characterological shorthand" for the characters' involvement in the systems and structures of American power ("Hunter" 189). Later, in his influential essay "Postmodernism, or the Cultural Logic of Late Capitalism," Jameson would use similar terms to indict *all* postmodern arts—science fiction and other "entertainment literature" in particular—for having made technology into a "privileged representational shorthand" for grasping complex social and political realities that are "beyond the capacity of the normal reading mind" (79, 80). Such narratives, in Jameson's view, indicate a "degraded attempt—through the figuration of advanced technology—to think the impossible totality of the contemporary world system" (80). Jameson assumes, however, that such a system (identified as nothing less than the "world system of present-day multinational capitalism" [79]) exists apart from the technologies by which it is registered. Never is he willing (no more than was Gerald Graff) to allow that current technologies such as the computer or television might have within them as great a capacity for representing contemporary "networks of power and control" as older technologies—railroad engines, turbines, "Sheeler's grain elevators [and] smokestacks"—had for writers and visual artists of the thirties and forties. The new technologies certainly are, as Jameson points out, "machines of reproduction rather than of production," but that in itself should not frustrate representation, though it may well channel representation into more abstract, difficult, and self-conscious modes ("Postmodernism" 79).

I have chosen these words carefully to describe not only the art of *Vietnam* but also works by American novelists and literary theorists

6. Here Jameson anticipates Thomas Schaub's much more thoroughgoing critique of Mailer's ideological complicities with corporate capitalism and contemporary "consensus" liberalism; see chapter 1 of this volume and, especially, Schaub's discussion of Mailer's cult of imagination and the "language of capital acquisition" in "The White Negro" (*Cold War* 158).

that *Vietnam* anticipates in its linguistic ambition, in its abstraction, and in its combination of themes that are at once occult and "scientific." *Vietnam,* though less encompassing than the novels of Thomas Pynchon, Joseph McElroy, or Don DeLillo, easily surpasses the "faulty representations of some immense communicational and computer network" which Jameson finds characteristic of most postmodern writing ("Postmodernism" 79). D.J. is certainly mesmerized and fascinated by technology, but the "occult element" within his narrative does not necessarily separate him from actual relations of power in the world. His impersonal representations and abstract connections imply social configurations beyond simple competition, and to some extent beyond the vantage of the social and psychological self.

The dream field or M.E.F. is one such abstraction, part of an animated landscape that, as Jameson notes, takes to itself all the unclean dreams and unconscious impulses of the North American continent. Yet, like the dream in all of Mailer's writing, the M.E.F. contains not only the intense masculine competition within American capitalism but also the possibility of communication among many separate beings, human and nonhuman, to discover "the meaning of trees and forest all in dominion to one another and messages across the continent on the wave of their branches . . . some speechless electric gathering of woe" (*Vietnam* 196). This is not simply the American nostalgia for a vanishing wilderness that Mailer famously shares, as a self-conscious if somewhat parodic equal, with Cooper in *The Pioneers,* Twain in *Huckleberry Finn,* and Faulkner in *The Bear.* The freedom D.J. and the other hunters experience in the Alaskan wilderness may well be the escapist liberty of men who are overwhelmed by giant forces of organized control. Yet Mailer does not suggest that salvation can be had by abandoning our technological existence. It is Mailer's point that there can be no clear separation between the technological society and the "communes of spookiness, pales, dominions, psycho swingers—even telepathies" that D.J. perceives in northern Alaska (159).

Having already referred more than once to the "general fission of the psychomagnetic field" (which was now "a mosaic, a fragmented vase as Horace said to Ovid" [114]), Mailer gets around to defining the dream, with mock scientific accuracy, in Intro Beep 10. In sleep, D.J. tells us,

> there's only one place you go, and that's into the undiscovered magnetic-electro fief of the dream, which is opposed to the electro-

> magnetic field of the earth just as properly as the square root of minus one is opposed to one. Right! They never figured out yet whether light is wave, corpuscle, or hung up on finding her own identity, all they know once you get down to it is that light is bright, and therefore not necessarily opposed to being part of Universal Mind. (*Vietnam* 170)

"All the messages of North America go up to the Brooks Range . . . crystal receiver of the continent" (*Vietnam* 172). But if the sum of these messages reduces in D.J.'s consciousness to a single purposive embodiment, Mailer is himself sensitive to diversity and complementarity in the dream field. Uncharacteristically for him, he even locates within science a conceptual model for achieving a style that might militate against its own universalizing and control-oriented tendencies.

Such a style is expressed partly by the idea of complementarity itself—that neither a wave nor a particle interpretation can fully comprehend the reality of light. More than a specialized theory of the physical world, this ability to live with paradox and contradiction implies a general habit of mind, whose literary application can be felt, in the passage just quoted, in the humor and the ironic attitude with which the whole idea of "Universal Mind" is presented. For it is through the sustained humor of D.J.'s style—a humor and self-irony that I am afraid is mostly absent from *Dream*, *Evenings*, *Song*, and *Harlot's Ghost*—that Mailer allows us to see the absurdity of our condition, just as this same humor enables Mailer himself to break up his own favorite metaphor of a doomed and dualistic American psyche and even, in natural descriptions of exceptional power and beauty, to go beyond his beloved "dialectic."

This willingness to unsettle his own formal arrangements, stylistic habits, and preferred metaphors has been appreciated as a primary source of value in Mailer, one so strong as to make acceptable a certain formlessness in even his most inspired writing. At the deepest level, the simultaneous building up and breaking down of literary form suggests the imaginative writer's most hopeful response to technology. Poirier, recognizing that a style wholly opposed to technology is bound to rigidify in devices of its own, finds technological affinities everywhere in Mailer—in the imperative he feels to innovate, to produce ever new and "improved" forms, to work contextual changes and constant repetitions that have the "stylistic capacity to match the tempo of historical accelerations toward disaster" (*Mailer* 131). Rather than quixotically *opposing* this acceleration, Mailer in *Vietnam* enters the field of technological operations as a self-aware, ironic, and non-innocent participant. He gives up any attempt at organic embodiment and seeks a subject position that is abstract, self-parodying, and unfixed to a reified mind or body.

Such a subject position is borne out in the book's last chapter, which avoids closure even as Vietnam is revealed as the inevitable end to which the entire narrative has been leading. The dream (M.E.F.) dissolves at dawn and becomes its complement (e.m.f.), and in an instant D.J. takes us from Alaska "into the new life smack right up here two years later in my consciousness, D.J. here at this grope dinner in the Dallas ass manse, given in my honor, D.J., I thank you, because tomorrow Tex and me, we're off to see the wizard in Vietnam. Unless, that is, I'm a black-ass cripple Spade and sending from Harlem. You never know. You never know what vision has been humping you through the night" (*Vietnam* 207–8).

This last possibility, that the narrator is not a late adolescent from Dallas, Texas, but a Harlem black masquerading as a Texan, represents one of the most problematic—and most significant—aspects of the novel. Initially put forward in the second Intro Beep and mentioned sporadically thereafter, D.J.'s alternate, "shadow" identity is never developed. Mailer makes no attempt to characterize D.J.'s alternative self but merely proposes it as a possibility, so we have no way of accommodating the black narrator within any of the terms the book has made available (unless, as with Shago Martin in *Dream*, we view him as an attempt by Mailer to appropriate the voice and gestures of a minority culture). In the end we are left only with one unanswerable question and a sign-off: "Which D.J. white or black could possibly be worse of a genius if Harlem or Dallas is guiding the other, and who knows which? This is D.J., Disc Jockey to America turning off. Vietnam, hot damn" (208).

In the drafts Mailer tried out a number of alternative endings—including an ending in which the narrator is revealed as an Italian, one "Norman DiMale"—before settling on this one. There is a pencil scrawl on the back of the title page in which D.J.'s double is revealed as not just any black from Harlem: "You see I wasn't killed in *An American Dream* like [old][7] Rojack think, shit no, this is Shago, Shago Martin, King of *An American Dream*, and this has been my black-ass fantasy about the white people I fear the most—Dallassassins—from . . . Texas." Shago's

7. "Old" is crossed out in the manuscript.

composite characterization as an "Elizabethan chorus" of voices is an attractive alternative to the "easy harmonious concordium of voice" that Rusty Jethroe's yes-men and corporate "assholes" present to their employers, like the creditors in *Timon of Athens* who answer "in a joint and corporate voice" (2.2.206). But the impulse to locate a style of political opposition in the black is one that, I think, Mailer was right to reject. Rather than again attempt, as he had done in *Dream*, to appropriate Shago's voice and subject position for his own self-critique—rather than absorb all otherness into the single, all-consuming consciousness of the egotistical sublime—he does better to recognize in himself political, racial, and technological complicities, and to discover a style of resistance *within* the culture he opposes, in a schizophrenic English that mixes elements from every locale in the culture.

D.J. in all his appropriating zeal is made to participate in a wider conceptual field; his subjectivity remains split and self-contradictory, disrupting any position of wholeness or comfortable category in which the reader might rest. Almost despite himself Mailer would seem to have created in D.J. a subjectivity that resembles nothing so much as Donna Haraway's feminist "cyborg," a part-human, part-technological embodiment whose subjectivity is "partial in all its guises, never finished, whole, simply there and original; it is always constructed and stitched together imperfectly, and *therefore* able to join with another, to see together without claiming to be another" ("Postscript" 22). Monstrous and violent though he may be, and despite all his claims to be a Harlem black, D.J. remains closer to those hopeful or promising "monsters" that Haraway celebrates, boundary creatures who in their evolutionary ability to contain contradictions and to adapt to a changing social environment may themselves come to be "signs of possible worlds . . . for which 'we' are responsible" ("Postscript" 22). The sensibility D.J. embodies is one that is very much in evidence in some of the most adventurous contemporary writing, by scientifically informed theorists such as Haraway no less than the imaginative writers she celebrates.[8] But the postmodern irony that Mailer momentarily antici-

8. Samuel Delany, Octavia Butler, Joanna Russ, and Vonda McIntyre are among the science fiction writers Haraway mentions. The celebrated cyberpunk fiction of William Gibson and Bruce Sterling, and literary precursors such as Pynchon's *Gravity's Rainbow*, Joseph McElroy's *Plus*, and Russell Hoban's *Riddley Walker*, independently create forms of technological consciousness analogous to Haraway's cyborg. In his acclaimed 1990 novel *Hopeful Monsters*, Nicholas Mosley develops biological metaphors similar to Haraway's. See my discussion in chapter 4; and on McElroy's *Plus*, see chapters 5 and 6.

pated in Haraway proved ultimately inconsistent with his commitment to an egotistical, dynamic version of the sublime, and this commitment in turn prevented him from taking complementarity seriously as an alternative to a fiction grounded in endless conceptual oppositions.

Pynchon, as we are about to see, does take complementarity seriously, and it is not the least of his accomplishments in *Gravity's Rainbow* to have been able to write ironically without sacrificing the more capacious gestures of the sublime. But irony alone cannot produce the hybrid, other-regarding consciousness that Haraway calls for in her "Cyborg Manifesto." Irony may well be an effective means, as Mailer discovered in *Vietnam*, of annihilating universalizing tendencies in oneself and in the technological culture. But when that culture is itself radically ungrounded, and when even the great monolith of science has given up its claims to universality and epistemological certainty, irony ceases to have a real oppositional force. Since neither Pynchon nor Haraway is as committed as Mailer to a rhetoric of opposition, they are perhaps less concerned with the limitations of postmodern irony. Their particular postmodern version of the sublime is certainly less focused on the individual ego, and irony in their work forestalls any of the ego's grosser ambitions for transcendence or its more regressive resistances to alienation of any sort. But before even a cyborg consciousness can join with another, disabused though it may be of romantic illusions about two essences merging into a single primordial essence, irony may yet have to be put aside.

3

Meteors of Style: *Gravity's Rainbow*

Let Friedrich Kittler, the German romanticist turned postmodern media theorist, introduce us to the world Pynchon writes about:

> *Optical fiber networks.* Soon people will be connected to a communication channel which can be used for any kind of media—for the first time in history or for the end of history. When films, music, phone calls, and texts are able to reach the individual household via optical fiber cables, the previously separate media of television, radio, telephone, and mail will become a single medium, standardized according to transmission frequency and bit format. Above all, the optoelectronic channel will be immunized against disturbances that might randomize the beautiful patterns of bits behind the images and sounds. Immunized, that is, against the bomb. For it is well known that nuclear explosions may send a high intensity electromagnetic pulse through traditional copper cables and cripple the connected computer network. (101)

The fiber-optic technologies Kittler advertises were of course still a way off when *Gravity's Rainbow* appeared. By 1973 the Pentagon was only beginning to assemble "the ARPAnet and various other radio and satellite networks" into a protected global internet, with computer destinations purposely decentered in order to withstand bomb attacks (Krol 13). But Pynchon surely perceived the link between global communication and the bomb, and realized that the fulfillment of either one could mean an end to history. "Critical mass cannot be ignored," he has one character say in a lecture. "Once the technical means of control have reached a certain size, a certain degree of *being connected*, the chances for freedom are over for good. [. . .] It is possible that They will not die. That it is now within the state of Their art to go on forever—though we, of course, will keep on dying as we always have" (*Gravity's Rainbow* 539).[1]

This lecture comes to us broadcast, appropriately enough, over remote monitors at a multimedia conference in hell, during one of Father Rapier's celebrated "critical masses," a pun on the critical mass of uranium needed to fuel the new "Cosmic Bomb" (539). The religious overtones are not accidental. Technology in this passage is more than a way of looking at the world which secular readers can believe in: it presents itself as a full-fledged version of a theological sublime, an apotheosis of human reasoning in a single synthetic Word. Yet, the structure of sublime transcendence (in Pynchon as in those English romantic poets studied by Thomas Weiskel in *The Romantic Sublime*) is as much rhetorical as it is theological. The rocket, worshiped by many characters in *Gravity's Rainbow* as a "Holy Text," provides a sublime uplift *as* text, a disembodied web of information that floats above nature's gravity and belies its potential for causing real, material destruction. Language conceived as universal and all-powerful, the apocalyptic convergence of word and world, the belief that we can substitute for nature an image of our own complexity—these are wholly secular replacements that inspire if not the reverence of a religious deity, then a new kind of dread and anxiety in the presence of vast technological systems.

Interestingly, Father Rapier's argument in *Gravity's Rainbow* has been cited and elaborated on with a remarkable literalness by cyberpunk writers such as William Gibson and Kathy Acker. This particular passage has lived beyond the cold war era not because of the bomb ("whose cousins are rusting in their silos all over Siberia or the Dakotas" [Siemion, "Whale Songs" 255]), but because Rapier's vision of an unlimited yet complete connectivity speaks to the virtual world these younger writers inhabit. In *Gravity's Rainbow* Pynchon intuited what people can know directly today: that the concrete materials of technology, even in its most destructive forms, are primarily epiphenomena, "signs and symptoms" of a reality that is abstract (159). Kittler's indestructible array of beautiful bits constitutes the apparently imma-

1. Because of Pynchon's frequent use of ellipses, my own are enclosed within brackets in all quotations from his work.

terial space of domination beyond individual comprehension (as the sheer scale of death by atomic war has been beyond the power of art to represent). Yet these abstract processes now reach the "individual household" with ease, creating an altogether different and unexpected relation to the technological life-world—the "technosphere," as it is now called. The overwhelming dread of the one great unimaginable event would seem to have been replaced, in the work of Pynchon's cyberpunk successors, and in Pynchon's own writing after *Gravity's Rainbow*, with a series of distractions, minor shocks, and endless stimulations of consumer desire.

It is only natural that readers should want to turn such images' total domination into something positive. Critics have thus celebrated connectedness in the postmodern novel as a source of dreams and mythic depths, limitlessness as a ground for textual indeterminacy and a seemingly infinite profusion of self-consciously textual details. Critics of Pynchon in particular have identified the proliferation of absent centers in his novels as a saving irony, a postmodern form of pleasureful play free at last from the stony-faced gravity of a totality that is well lost. But are these textual features really opposed to technologics of domination and control? In the absence of former certainties that might sustain a collective reality, many of Pynchon's characters project worlds of their own, in anticipation of the unconstrained projections of private desire that now characterize the virtual world of computer simulation technology. In both cases, the literary and the technological, to project a world is to indulge a wholly *linguistic* freedom, an excess and autonomy of the signifier loosened from the necessity of connection with the world outside the projecting self. The prospect of such freedom would no doubt explain why contemporaries of Pynchon trained in an academic version of poststructuralism have placed so high a value on irony and indeterminacy, but it is a freedom that the narrator of *Gravity's Rainbow* might well reject as falsely "comforting," a weak "anti-paranoia" to the "paranoia" of virtual worlds and totally connected systems (434).

"Shall I project a world?" Oedipa asks in *The Crying of Lot 49* (82). Oedipa, and Steven Rojack and William Cartwright, are remarkable characters in postmodern fiction in that they actually put their illusions to the test and work to verify the coherence they read in events. Oedipa tries to determine whether the connections she makes are sustaining dreams or phantoms that dissipate like the self-generated and self-encompassing images in a virtual reality display. Illusions are not

always *de*lusions, especially not when we can go back to them later, as we might go back to a written text, and compare their reality with our ongoing experience of the world. But the trouble with most projected worlds is that, in the absence of such referential testing, they turn out to be mere *surface* projections from the mind's arbitrary center which only the paranoid would mistake for reality.

Pynchon likes to present characters in mental states that fluctuate between the total theory of a paranoid delusion and the ironical "mindless pleasures" of a total relativism. The overdetermined and wholly private meanings in the first state of mind are dissolved in the second by an irony that would undermine the ground on which any stable meaning might be built. In the end, however, neither state receives authorial sanction, for neither one does anything to advance the radical freedom that clearly concerns Pynchon, however much it eludes his characters. Neither the straight (paranoid) nor the ironical (antiparanoid) projection does anything at all to resist whatever power, totalitarian or otherwise, happens to be in place at a given moment. Postmodern irony *changes nothing* (and not in the potentially subversive way that philosophy, for Wittgenstein, "changes nothing" except our conceptual orientation toward the ideological status quo, which leaves open the potential for us to change *everything*). The elaboration of an ironic, self-consciously linguistic universe is the problem in Pynchon, not the solution to fragmentation, uncertainty, and alienation in the technological subject, as novelists and critics of the generation after Pynchon are coming to realize.

David Foster Wallace is perhaps the first writer of this younger generation to have perceived just how "poisonous" postmodern irony has become: "If I have a real enemy," Wallace told an interviewer, "a patriarch for my patricide, it's probably Barth and Coover and Burroughs, even Nabokov and Pynchon. Because, even though their self-consciousness and irony and anarchism served valuable purposes, were indispensable for their times, their aesthetic's absorption by U.S. commercial culture has had appalling consequences for writers and everyone else" (McCaffery, interview 146).[2] There is no more pointed fictional meditation on commercial culture's cooptation of high post-

2. Here Wallace is characterizing a trendy postmodernism that includes the contemporary writer Mark Leyner, author of *My Cousin, My Gastroenterologist,* as well as television and radio personalities from David Letterman and Gary Shandling on down to Rush Limbaugh, "who may well be the anti-Christ" (146–47; presumably Wallace is here being ironic himself).

modernism than Wallace's brilliant story "Westward the Course of Empire Takes Its Way," which has John Barth (named "Ambrose" after his most famous fictional persona) collaborating with the advertising director of McDonald's restaurants on a nationwide chain of "Funhouse" discotheques. Ambrose, who, like Barth, teaches creative writing at a prestigious university in the Tidewater region of Maryland, came of age as a writer and academic when New Criticism held sway. For all his postmodern funhouse complexities, Ambrose/Barth upholds the new critical value of irony as a determining criterion of literary structure and excellence, a formal unity that comes through, and may even be protected by, all the ironic funhouse antics. But few academics have understood the uses of such irony to mask, or smuggle in, not just a literary complexity adequate to human feeling, but an absolute and elitist knowledge of reality. In a postmodern culture where the only absolute value is determined by world markets, irony and indeterminacy (in advertising and television, and even in Corn Belt politics) become powerful legitimations, ways of adjusting to the economic absolute while upholding the appearance of hip radicalism.

It is not surprising that Pynchon's linguistic pyrotechnics should have caused critical contemporaries who suffered under the New Critical tutelage to miss the deeper, more radical freedom of *Gravity's Rainbow*, which strangely combines extremes of indeterminacy and interconnection, postmodern irony and the greater expansiveness of the romantic sublime. Pynchon's sublime does not construct a world system but subverts the status quo of *all* systems, be they scientific, linguistic, or ideologically inscribed as "human." His sublime, in other words, is an eminent postmodern instance of Weiskel's "negative" or "critical" sublime, whose intensity is a function not of a recuperated rational system or of an ironic dependence on the very systems it purports to reject, but of "the impossibility of realizing (in any way) the idea of humanity (or any supersensible idea)" in language (Weiskel 45). There is always in Pynchon a semiotic shadow, a "delta t" that falls between the inscribed word and "the thing it stands for" (*Gravity's Rainbow* 510); and this un-ironic critical distance is necessary in his writing if he is to avoid the apocalyptic collapse of word into world.

Of the four writers treated at length in this book, Pynchon is the one who has proved capable of releasing the full un-ironical force of the sublime as a *stylistic* gambit, to the extent that he risks indulging in the "meteors of stile or False Sublimity" that the German critic Samuel Werenfels isolated for censure in his eighteenth-century contemporaries. Without doubt, in Pynchon's work can be found every instance of bad writing that Werenfels lists: a solipsistic obscurantism, the use of "Curios" of language that substitute "hard words" for common terms, absurd or ludicrous tropes, the lavishing of a high style on unworthy subjects, and an overall "excess of elevation without 'Decency and Moderation'" (see Edward Tomarken's introduction to Werenfels vii–viii).

Admittedly, it would be tendentious to maintain such old standards for a postmodern writer like Pynchon. But the issue here is not primarily one of literary quality. In setting out these stylistic excesses and infelicities, Werenfels and more recent critics of the sublime are getting at something deeper, namely, a nonreferential quality that threatens to undermine sublime writing, the specious supposition that "language may create a reality unto itself, equal to, if not surpassing, that of the world of things" (Werenfels xi). Such a "reality unto itself," I am suggesting, is the linguistic counterpart to the self-validating technological systems described by Kittler, the cyberpunks, Jean Baudrillard, and the virtual reality engineers who project the worlds these writers theorize. It requires a stretch of the imagination to link linguistic and technological systems, but it is less of a stretch today, when technological systems have begun to take on a textual form in potentially infinite internets and hypertexts. These are all examples of an illusionary space (a virtual space) that promises something without limits and is yet a totality, a wall of enhanced sensation that does not really let the user touch anything, but that gives an *illusion* of total control over the environment. In virtual reality connections are generated between thought and the world, but all of the connections are self-created, channeled back through the senses, and still experienced as Other. The danger in all such self-validating systems is that they develop any which way, without internal resistances or any need for external, referential constraints. In such systems only technical breakdown, user fatigue, or boredom can halt the free play of signification, and (in a Baudrillardian extrapolation) only death can ensure the return to material reference that the sublime requires. That is a large extrapolation, to be sure, but it is hardly surprising that death is so often imaged forth, in what some have called a "nuclear" sublime characteristic of cold war fiction, by the bomb.

Is the bomb, then, the only referential reality that guarantees meaning in *Gravity's Rainbow?* At its best Pynchon's narrative conducts the reader along a projectile-like movement of thought, creating the heady illusion that the world is nothing more than a linguistic construct. In the absence of total systems, the rhetorical imagination substitutes its own free play of signification, where accidents of individual experience and historical, documentary realities accumulate and eventually take on the density of a substantial world. *Gravity's Rainbow* is not short on verisimilitude. Neither are the virtual realities that Pynchon's fictions anticipate. In both cases the mind (of a reader, of the person wearing a virtual reality headset) ends up substituting an image of itself and its own desires for any externally verifiable reality, and in the absence of material resistances the mind is "free" to experience the world in any way—that is, so long as the system doesn't crash and the bomb doesn't drop.

The rocket trajectory, that pure parabola in the sky to which "everything, always, collectively, had been moving" (209), is the novel's primary figure for an all-encompassing totality, the form against which all else is excess and unpresentable, lost to history. In following this trajectory the rocket describes a power that is universal, transcendent, and immediately transferable over great distances. But the rocket is not wholly reducible to this metaphysical symbol. The trajectory is mathematics, pure and transcendent, but the rocket is engineering; first and foremost it is "raw hardware," a concrete assemblage of parts and functions whose details are too profuse, and too firmly rooted in technological fact, to have been introduced solely as a contribution to a cultural mystique or mythological cosmos.[3] Which is to say that, unlike the trajectory, the rocket *needs* excess for its power. And Pynchon as a writer relies on excess over and above all symbolic meanings in order to root his sublime aesthetic in reality.

Using these distinctly nontranscendent, even dull details of space rocketry, readers of *Gravity's Rainbow* might piece together a different, less threatening image of technology. The rocket becomes neither a fetish nor a nightmare projection over the head of every reader, but a source of material resistances that are necessary both to the production and common experiencing of a human reality. In pursuing such engineering details, I find Pynchon to be instructively ambivalent about using the intellectual force of his language either to carry the reader to

3. We know that Pynchon during the composition of *Gravity's Rainbow* drew extensively on the standard histories and memoirs of working scientists and engineers: Kooy and Uytenbogaart on ballistics; von Kármán on aerodynamics; Sasuly on IG Farben; Wiener on cybernetics; Dornberger, von Braun, and Ley on the history of space rocketry (see Weisenburger for bibliographical details). That the novel is extraordinarily well researched even Pynchon's detractors must concede—if only to stigmatize him as an academic writer.

exalted heights or to penetrate "hidden" levels of the human psyche as it is embodied in the machine. Unlike Mailer and other contemporary avatars of the "positive" egotistical sublime, Pynchon avoids giving the impression that one could ever arrive at a "deep" or psychological meaning directly, at a moment of metaphorical displacement and condensation in which the object world is joined to the substance of mind. Agency in *Gravity's Rainbow* is not essentially self-projective, grounded in the ego which casts its own anxieties and ambitions onto the technological object. Mind and world, literature and technology in Pynchon remain distinct but homologous discourses, not to be joined (as we have seen Mailer unsuccessfully attempt to join them) by the most energetic acts of metaphor, and not to be internalized by the subtlest of psychological systems. Instead of setting the self in a relation of desperate opposition to technology, Pynchon affirms technology in its irreducible otherness, and he uses the language of science and engineering—as well as more traditional, melodramatic modes—to get outside the self.

Perhaps the most conventional critical response to technological material that so clearly resists integration into any total explanatory framework has been to invoke concepts of indeterminacy in modern physics and mathematics. It is true that Pynchon, like Henry Adams before him, is adept at discovering contradictions, and even occult mechanisms, within engineering methods and apparatus, and at seeking within the known universe of scientific forces data that hint at the existence of an unknown "supersensual" multiverse (381 passim). Yet the organization of complex technological imagery in *Gravity's Rainbow*—particularly the rocket trajectory and the human lives it embraces—expresses its own style of indeterminacy, a way of being that may be nearer in its details to engineering technology than to what are in fact quite precisely delineated indeterminacies of modern physics. Unlike physics, the field of engineering suggests no comprehensive *theory* of incompleteness or indeterminacy, but it does tend to ensure the *experience* of indeterminacy in those who practice it, if only because the engineer must continually respond to material contingencies in the realization of a design.[4]

The rocket's sheer destructive power, the repeated test runs and simulations, and the ever-present chances for breakdown offer a num-

4. The principles of complementarity and indeterminacy, and the general "field view" of modern physics, are discussed in chapter 4.

ber of possibilities to the literary imagination, suggesting ways of reading Pynchon that to me seem less restrictive than the usual emphases on entropy, paranoia, and apocalyptic science. Not only does Pynchon dramatize the engineer's experience as well as has been done in fiction, but he writes with much the same ready, unstructured sensibility, introducing scientific and technical metaphors with an eye to particular effects and solving unforeseen problems as they arise in the emerging design of the narrative.

Engineering in the early years of rocket research, with its unplanned responses to political and material contingencies, is both a subject and a method of *Gravity's Rainbow,* however much Pynchon (a former engineering student at Cornell University and technical writer at Boeing Aircraft in Seattle) is aware of the less creative, more bureaucratic aspects of the field. Engineering is also, in a novel so concerned with power and mastery in all its forms, a possible way of viewing Pynchon's own attempt to master the material of the never definitive "Text" of the rocket.[5] And it is here, I would suggest—in representing the ways power is exerted over and through instruments, apparatus, and the verbal material of language—that *Gravity's Rainbow* can reveal unnoticed affinities between technology and literary production.

Post-Terminal Trajectories

We can grasp the general contours of Pynchon's engineering psychology at the start, in the book's opening epigraph from Wernher von Braun expressing the rocket engineer's conviction that the lessons of science ensure "the continuity of our spiritual existence after death" (*Gravity's Rainbow* 1). Responses to this epigraph have ranged from one critic's measured acceptance of von Braun as a spokesman for the worldview of modern physics (Nadeau 138) to another's total rejection of the sentiment as a "cover-up" or rationalization for von Braun's own role in creating the V-2 rocket (Black 23). However the epigraph is taken, what is important is that such an assertion should have been made *by* von Braun, if only because it demonstrates that a concern with survival after death—even a dream of immortality—is not alien to scientific thought. By having von Braun announce his theme, Pynchon forestalls any assumption that his own fictional transactions with the

5. Tom LeClair discusses mastery as both theme and method of *Gravity's Rainbow* (*Art of Excess* 36–68).

Other Side are necessarily separate from the professional interests and psychology of scientists and engineers.

Within the fiction, as if he were taking von Braun at his word, Pynchon imagines a fully installed "bureaucracy" of the Other Side, whose spirits possess as much authority as do any of the book's living scientists. One so eminent as Walter Rathenau, who was "prophet and architect of the cartelized state" on This Side, is now, during a séance, telling the German technocrats who are his successors that all they "believe real is illusion" (164, 165). More than once Rathenau criticizes their belief in secular history as a progression through discernible stages ("You are constrained, over there, to follow it in time, one step after another. But here it's possible to see the whole shape at once" [165]), and his view of IG Farben, the cartel he helped to found, is equally heretical ("The more dynamic it seems to you, the more deep and dead, in reality, it grows" [167]).

Yet, there is another aspect to Rathenau, who has so far been, despite his superiority, entirely consistent with the capitalist's dream of unlimited wealth and power. In asking the elite from the corporate Nazi crowd to consider the "real nature" of synthesis and control, Rathenau is looking beyond the industrial process these men think they understand into "the technology of these matters. Even into the hearts of certain molecules—it is they after all which dictate temperatures, pressures, rates of flow, costs, profits, the shapes of towers . . ." (167). This, finally, is the crux of Rathenau's (and Pynchon's) message: there is an order in nature that determines the possible uses we can make of it, and beyond which no social, political, or technological order can go.

Rathenau's séance is typical of the way Pynchon uses the Other Side, not to contradict the laws of nature but to bring them more forcefully to bear against any dream of transcendence or unconstrained progress through technology. For all their differences of class, nationality, and temperament, Pynchon's technocrats and engineers have an identifiable psychology. Being expert in techniques that can transform dead matter into plastics, coal-tar dyes, and steel, and hoping, at times consciously, ultimately to free themselves from organic cycles of life and death, the engineers end by contributing to the design and manufacture of objects that accelerate the deathward movement within contemporary history. (The rocket is only the most obvious example.) As Rathenau's comments on the conversion of coal into steel make evident, the industrial process is essentially at one with psychological repression:

"Imagine coal, down in the earth, dead black, no light, the very substance of death. Death ancient, prehistoric, species *we will never see again*. Growing older, blacker, deeper, in layers of perpetual night. Above ground, the steel rolls out fiery, bright. But to make steel, the coal tars, darker and heavier, must be taken from the original coal. Earth's excrement, purged out for the ennoblement of shining steel. Passed over.

"We thought of this as an industrial process. It was more. We passed over the coal-tars. A thousand different molecules waited in the preterite dung. This is one meaning of mauve, the first new color on Earth, leaping to Earth's light from its grave miles and aeons below. There is the other meaning . . . the succession . . . I can't see that far yet . . . ." (166)

As Pynchon has Rathenau describe them here, the rainbow of synthetic colors contained within the coal-tars is the transcendent signifier that rises above the controlling "industrial process"—a transcendence that the engineers themselves are not aware of. (In the early 1930s, they could not have been aware of synthetic coal-tar dyes: only later would their discovery prove "crucial to the German munitions industry" [Hayles and Eiser 5]). But this passage looks beyond any single engineering innovation to a "succession" that even Rathenau cannot yet see. At this stage in industrial culture, it was still possible to distinguish "between a material production . . . and a production of signs," as is common in traditional Marxist analyses (Baudrillard, *Critique* 143). At the time when Pynchon was writing, however, we were already well into a post-industrial culture in which signs and exchange value had taken on a greater importance than the "industrial process." This transformation would also have an effect on human psychology, for, as Hanjo Berressem notes (summarizing Baudrillard), "the abolition of the real implies the abolition of repression as well; what remains is merely the *simulation* of repression" (Berressem 44).

But this would come later; the scientists, engineers, and corporation managers in *Gravity's Rainbow* are still within a repressive ideology of production and control. "Passing over" the dead organic matter buried in the earth, these men train their aspirations on the sky and the stars. They would perfect through technology an "Apollonian form," which is, in the words of Norman O. Brown, "form negating matter, immortal form; that is to say, by the irony that overtakes all flight from death, deathly form" (157).[6] In Brown's psychohistorical reading of the Greek

6. For a detailed treatment of Brown's presence in *Gravity's Rainbow*, see Wolfley.

myth, it was Apollo, the sun god and charioteer, "who gave man a head sublime and told him to look at the stars." Though Apollo is masculine, his masculinity "is the symbolical (or negative) masculinity of spirituality," which displaces the baser, instinctual yearnings "from below upward." The Apollonian world, Brown continues,

> is the world of sunlight, not as nature symbol but as a sexual symbol of sublimation and of that sunlike eye which perceives but does not taste, which always keeps a distance, like Apollo himself, the Far-Darter. And, as Nietzsche divined, the stuff of which the Apollonian world is made is the dream. Apollo rules over the fair world of appearance as a projection of the inner world of fantasy; and the limit which he must observe, "that delicate boundary which the dream-picture must not overstep," is the boundary of repression separating the dream from instinctual reality. (157)

What is of course dangerous about such a strict observation of limits is the way it can distance the engineers not only from their own sexual, instinctual reality but also from the synthetic, simulated dreamworld they help to create. Nor is this danger necessarily lessened by Pynchon's knowing that he, as a writer, must also employ controlling structures as well as various kinds of simulation or imaginative synthesis of social, technological, and historical materials in order to create a plausible fiction (i.e., one that constantly tests itself against a developing reality without yielding to reality's formlessness). Even repression and a degree of sublimation are necessary for a text to organize itself, and for the imagination to convey the idea of a coherent and knowable reality beyond the "fair world of appearance" that the novelistic imagination, like Apollo, must construct out of the reality it has internalized. Nietzsche's Apollo fashions a world of appearances which he perceives "mediately, in the contemplation of the plastic image" (Deleuze 12). To him belongs the contemporary simulation culture based on the perfection of technique, a culture whose direct presentation is neither tragic nor sublime but—at best—beautiful. The Apollonian world, in other words, is a transcendent fiction empowered wholly by technical simulations; its appearances are ultimately incapable of reflecting any reality outside itself.

The series of individual dream episodes to which I now turn, each one demonstrating ever more brilliant innovations in literary technique, finds a deep congruence with the administrative and engineering activity that is the book's primary subject. In themselves these innovations might also be taken for mere "beautiful appearances" which Pynchon "embroiders" onto the deeper, repressed reality that the civilized, technological mind has only recently (and partially) left behind—the realm of being that Nietzsche ascribes to Dionysus (Deleuze 12). But these episodes amount to more than a surface "projection" of the author's "inner world of fantasy." It is to Pynchon's credit that he pushes his own rational systems and psychic, suprarational conceits beyond the "delicate boundary" his characters may not even know they are observing, applying the pressure of a studiedly excessive language to the melodramatic depiction and eventual dismantling of the corporate dream. Meaning in *Gravity's Rainbow*, as in Nietzsche's "symptomological and semiological" science, is in the superabundance of production, in the real that exceeds symbolization (Deleuze 3).

Hence the prevalence in the book of interstitial images: intersecting worlds, merging orders, the "delta t" that posits a hitherto unimaginable mathematical function (division by zero) in order to produce an idea of infinity. This idea, in turn, allows one to extend a rational system beyond its inherent limits (making it possible, in this case, to calculate the area beneath a curve). It is typical of Pynchon's nonmetaphorical use of science that this last figure is wrested from its mathematical context and made to signify the distance separating "words" from "the things they stand for" (510). The mathematical symbol becomes, in Pynchon's hands, a *rhetorical* figure, a congruence that is based on no rational argument uniting the two fields but rather on a perception of a mutual vulnerability, of the necessary and fundamental limits of both the linguistic and the mathematical system. But it is important not to see this congruence as a mere celebration of irrationality, for the breakdown in both cases permits an *extension*, though on different terms, of a system's signifying power, an imaginative recovery after a system's collapse that is the very condition of the sublime.

Ronald Sukenick has observed that fictional examinations of technology often get into "causes rather than consequences, in a way surprisingly parallel to psychological analysis" (61). In *Gravity's Rainbow* Pynchon would seem to fulfill the psychoanalytical possibilities within modern technology by making its surface forms an expression of the latent content of the engineering mind. His method is melodramatic in the sense defined by Peter Brooks in *The Melodramatic Imagination*, a way of using excessive modes and "enactments" that "would, in their breakthrough of repression, carry the message of our inner selves" and would ultimately "[accede] to the latent through the signs of the world" (202). The idea is not to discover known causes, as one character in the novel attempts haltingly to tell her uncomprehending engineer husband, but to move *through* technological "signs and symptoms" to the unconscious and the unknown (*Gravity's Rainbow* 159). The signs of modern technology—its products, methods, and apparatus—are symptoms of forces that are latent or immanent in the world. Such immanence is not transcendent idealism in another, psychological form, for Pynchon never allows his connections to resolve themselves into a coherent system or to converge on a single, complete perception available to the ego. Nor should we think of this latent meaning as being "hidden" behind the form of the dream, as conventional psychoanalysis often conceives of a meaning "beneath" appearances that the form of the dream is meant to hide. Meanings in *Gravity's Rainbow* are not at all "deep" in this sense; they exist right on the surface, in the form and mechanisms of the dream-work itself.

If we shift from the depths of psychology to the surface model of semiotics, we might imagine Pynchon to be moving out beyond the boundaries that separate multiple sign systems into regions where new connections may be drawn and other meanings generated. Rather than complete a system of his own, he builds connections *between* systems. In presenting his technical material, he invents occult passages through complicated routes and switching paths in order to reveal operations within or behind the industrial process that escape even those characters who are most involved in it but are too locked into their own system to see beyond it. These characters often receive such information while experiencing literal dreams, hallucinations, or altered states of consciousness, such as the "one particular state of consciousness" that Tyrone Slothrop "find[s] his way into" during his course of instruction in V-2 flight dynamics at the Casino Hermann Goering (237). Slothrop himself is not aware of it, but he is in touch with the spirit of "the late Roland Feldspath [. . .] long-co-opted expert on control systems, guidance equations, feedback situations for this Aeronautical Establishment and that" (238). Feldspath (a path through a *Feld*, the German for "field") transmits along one of Pynchon's many rainbow arcs, in this case one of the final trajectories that might have been taken by the V-2 rocket itself had the Germans continued their strike through the spring of 1945 against advancing American troops in Europe. By thus stationing Feldspath "along one of the Last Parabolas—flight paths that must never be taken," Pynchon creates a route of information and potential destruction that is congruent with actual flight paths but at the same time separate from them, as if he were trying to emphasize that such excursions, though flights of the imagination only, might stand nonetheless in close relation to actuality (238).

There is no more extreme instance of Pynchon's excess and melodramatic expressionism than Feldspath's hypothesis that an entire generation of German engineers came to practice their profession as a reaction against the runaway inflation of the early 1920s. Feldspath in an earlier séance at "Snoxall's" had already applied his theories of control and guidance to economies, suggesting that a market "now could *create itself*—its own logic, momentum, style, from *inside*" (30). And so it might have been, if one is paranoid enough to make the connection, that "the whole German Inflation was created deliberately, simply to drive young enthusiasts of the Cybernetic Tradition into Control work: after all, an economy inflating, upward bound as a balloon, its own definition of Earth's surface drifting upward in value, uncontrolled, drifting with the days, the feedback system expected to maintain the value of the mark constant having, humiliatingly, failed . . ." (238).

It is doubtful that Pynchon intends us to take this altogether seriously as a reason for the "German Inflation." If an inflationary economy was in fact created deliberately, the probable reason had to do with war debts and reparations to France, England, and America. (Payments were made in terms of the gold standard, not the devalued mark, but it was in the interest of the Weimar government to claim that their cost was ruining the country; compare the endless references to the "cost of reunification" made by the Kohl government after the fall of the Berlin wall in 1989.) Also, it is too early to be speaking of a "Cybernetic Tradition" in Germany.[7] Yet one cannot discount Pynchon's insight into the psychological effect that an economy apparently subject to no external determinations must have had on those who came of age and had to earn a living during the years of uncontrolled inflation. Engineers in the novel such as Franz Pökler and Kurt Mondaugen would surely have found a certain satisfaction in control work, an escape of sorts from the social deprivations and senseless meanderings of daily life in a Berlin slum. For such men the engineering profession could be both an economic and a psychological refuge from the confusion produced by a symbolic system that has no reference to anything outside itself.[8]

7. Weisenburger (126) cites the summer of 1947 as the date when, according to Norbert Wiener, the term *cybernetics* first came into general use.

8. In *V.* Mondaugen suffers a "soul-depression" in Sudwestafrica that is analogous to the economic depression he lived through as a student in Berlin

It is this very desire for order and control in one's life and work that becomes, for Pynchon, the motivation and ground for larger, collective forms of control. Many commentators have mistakenly identified an impersonal, aloof, and sinister plot-making activity within the novel as the primary subject of Pynchon's political critique, a subject that can be made to seem only slightly less dubious for including his own construction of novelistic plots and strategies. In my view the paranoid plot-makings attributed to the catchall terms "Them" and "the Firm" are less successful, and generally less important, than are the more particular forms and methods of control that modern technology instills in us all but that are most concretely expressed in the novel in the working lives of scientists and engineers.

In the present passage Pynchon is most interesting not when he is proposing the doubtful existence of some sinister manipulating presence behind the German economy of the 1920s. There are more intimate and particular activities that predispose us to processes of social and political control, and it is Pynchon's unique accomplishment to have located and made such processes concrete in the material forms of engineering technology. It is by such means that the entire social, economic, and psychological situation of the engineer comes to be expressed through the single cybernetic operation of "feedback," which can be any mechanical or electronic reflex that lets a system adjust to external circumstances. It is feedback in the circuits of the V-2 rocket, for example, that keeps it moving tangent to its trajectory. As Pynchon describes it, making confident use of engineering argot, feedback in the rocket's guidance unit ensures "unity gain around the loop":

> Unity gain, zero change, and hush, that way, forever, these were the secret rhymes of the childhood of the Discipline of Control—secret and terrible, as the scarlet histories say. Diverging oscillations of any kind were nearly the Worst Threat. You could not pump the swings of these playgrounds higher than a certain angle from the vertical. Fights broke up quickly, with a smoothness that had not been long in coming. Rainy days never had much lightning or thunder to them, only a haughty glass grayness collecting in the lower parts, a monochrome overlook of valleys crammed with mossy deadfalls jabbing roots at heaven not entirely in malign playfulness. (238–39)

---

(277). Pökler's time out of work and living on half-rations in the 1930s is presented in *Gravity's Rainbow* (158–63).

The writing here, as so often in *Gravity's Rainbow,* is remarkable first for the ease with which a technical image comes to stand for both a state of mind in the young engineers and a corresponding external condition. The playground swing, which is Pynchon's—the former science writer's—homely analogy for more complex feedback systems, leads metonymically, sentence by sentence, into a description of an actual playground scene, becoming part of a colorless entropic landscape that serves as an outward symbol of the engineer's desire, in life as in work, to keep within definite limits: "Just as no swing could ever be thrust above a certain height, so, beyond a certain radius, the forest could be penetrated no further. A limit was always there to be brought to" (239). These are men of moderate and limited vision, although it turns out, for Pynchon as for Melville in *Benito Cereno,* that "the moderate man is the understrapper of the evil man."

The passage as a whole reveals an aspect of Pynchon's writing that is not often remarked, even by those who applaud his technological sophistication or by those more rare readers who have noticed how good he can be in scenes of natural description.[9] I am thinking of how the "monochrome overlook of valleys" and "mossy deadfalls" of this passage are not *opposed* to the governing technological metaphor, but seem rather a natural outgrowth of a scientific concept that might be used to describe human and natural process—the breakup of a fight, the duration of a rainstorm—as easily as it describes the rocket. Pynchon is certainly not the only contemporary writer to have perceived a subject-object rift within traditional technological ways of thinking; but where so many other writers attempt to paper over or cover up the rift with ornamental images of nature in its pretechnological wholeness, Pynchon tends to *work through* the systems and images created by engineers to a holistic vision that does not exclude the systems and images. Rather than set his own metaphors, speculations, and imaginative constructions in opposition to technological constructions, he returns to them with an obsessive fidelity, to the extent that he willingly risks a novelistic complicity with the very controlling structures and reductive techniques that characterize the thought and work and "deep conservatism" of the "young engineers" (239).

Consider his use of the rocket trajectory itself, which as a title image and indeterminate unifying motif is no less recurrent and mechanically presented than was the letter *V* in his first novel. Without losing

9. David Leverenz, almost alone among early critics of *Gravity's Rainbow,* mentions Pynchon's "aching tenderness for natural description" (244).

sight of actual flights and their historical employment in struggles for geopolitical power, he is able also to recognize and make literary use of the engineer's conceptualizing power by considering the trajectory abstractly, as a mathematical potentiality that can be applied in many different situations. Throughout the novel he depicts the engineer's professional progress as a "tightly steered passage" that keeps a company man such as Klaus Närrisch well within the bounds of moderation (517). The conceit is also evident in the way Franz Pökler is treated by his superiors at Peenemünde: "Any deviations into jealousy, metaphysics, vagueness would be picked up immediately: he would either be corrected back on course, or allowed to fall" (417). The rocket trajectory in such passages describes both a personal and a collective destiny, extending, as at the close of Slothrop's reverie, beyond the grave into either a void or an aether where "souls [. . .] Rocketlike" are driven "out toward the stone-blue lights of the Vacuum under a Control they cannot quite name . . . the illumination out here is surprisingly mild, mild as heavenly robes, a feeling of population and invisible force, fragments of 'voices,' glimpses into *another order of being . . .*" (239).

The perception of such masses and the confusion of so many voices is to a large extent what Pynchon, no less than any other modern writer, is trying to make sense of. And he is rightly credited with creating a dense, multivocal narrative equal to the complexity of a mass society. Yet his effort requires an active and even at times aggressive reduction of human lives to the terms of his own metaphors (a life, a history, is like a rocket trajectory), or he must displace the crowded voices into a half parodic spiritual realm or some other pre-capitalist myth.

A passage such as this one suggests that he too must employ the very techniques of our modern technological society to avoid being overwhelmed by it. That he may very well need such techniques in order to resist them with any effectiveness is evident in the more recent, more relaxed—and to this extent weaker—novel *Vineland*. There, the threat and oppressiveness of a multivocal society is again evident, and much more directly presented in all its cultural particularity. But this time Pynchon gives the discomforting vision to the control freak Brock Vond, whom he places in a cacophonous hell that reverberates with "the drumming, the voices, not chanting together but remembering, speculating, arguing, telling tales, uttering curses, singing songs, all the things voices do, but without ever allowing the briefest breath of silence. All these voices, forever" (*Vineland* 379). Although Paul

Maltby rightly identifies these voices as originating in the mythic and indigenous Yurok people and so suggesting "the enduring vitality of pre-capitalist narratives" (182) in America, he must ignore the voices' menacing aspect. In never "allowing the briefest breath of silence," are they any less overwhelming than the capitalist totality they supposedly resist?[10]

In noticing the ways in which Pynchon relies on controlling structures analogous to those he invokes, and admitting even the extent to which in *Gravity's Rainbow* he perhaps "willingly accepts, and accentuates, a writer's unavoidable complicity with the plots that torture his characters," we need not entertain the suspicion, with Leo Bersani in an extraordinary essay, that Pynchon as an author might be "one of Them" (188).[11] It may be impossible, it is true, to separate the restrictive forms within contemporary technologies of control from Pynchon's own activity as a form-maker in fiction; but this does not lessen the power of Pynchon's resistance, at least not in *Gravity's Rainbow*. Pynchon is doubtless aware of the limits to all merely formal arrangements, but he *depends on* the existence of such limits in order to create a sense of freedom at the point where they break down or, better yet, at the point where two or more separate rational systems come into conflict with each other, creating the spiritual equivalent of worlds intersecting. So far from repressing an awareness of his own rational systems, Pynchon accentuates their limits in order to extend their signifying power beyond themselves, into a noumenal world "between" powers, and perhaps even beyond the precincts of the controlling imagination.

Later in the novel, as if to explode any notion that his own or any

10. Pynchon is expressing, obliquely, much the same difficulty that Lynne Tillman articulates in a report from Ellis Island, at "the museum of hyphenated Americans." Like Pynchon in *Vineland*, Tillman discovers in herself a need to exert some *narrative* control over the "heteroglossia" of immigrant voices. In one exhibit, Tillman hears "the words of the immigrants themselves, on audiotape recordings that filled the rooms with a cacophony of differently accented voices. A novel's many voices, [Tillman's protagonist] Madame Realism thought again. In addition to their voices was Madame Realism's —she was speaking into her tape recorder, hoping to capture, with something like immediacy, some of what she was viewing and some of what was written in the many captions. She noticed that several of the other visitors found this peculiar, even suspicious, behavior. And seeing this she felt, in her own way, suspect. But wasn't she merely trying to be her own narrator?" (57).

11. I addressed Bersani's provocative hypothesis in a version of this chapter published in 1992, in *Novel*. I am now willing to entertain Piotr Siemion's premise that, in *Gravity's Rainbow*, "them" means *us* (see the chapter on Pynchon in "Whale Songs"). Siemion finds overoptimistic my conclusion that Pynchon's aesthetic is not circumscribed by the engineering systems it employs as narrative constraints.

historical narrative might do more than suggest the reality of the mass technological society he invokes, Pynchon's narrator populates the sky with an entire "constellation" of rocket trajectories, each parabola "an interface between one order of things and another. [. . .] They still hang up there, all of them, a constellation waiting to have a 13th sign of the Zodiac named for it . . . but they lie so close to Earth that from many places they can't be seen at all, and from different places inside the zone where they can be seen, they fall into completely different patterns . . ." (302).

So, too, are Pynchon's many supernatural digressions constellated about the text of *Gravity's Rainbow*, the entire constellation falling into any number of "different patterns," and each pattern, depending on where one views it, in turn suggesting a different interpretation. This is not the indeterminate and floating reality that has led many postmodern theorists to propose an end to history. The coexistence of conflicting interpretations in *Gravity's Rainbow* is hardly peaceful; surely such coexistence does not reduce Pynchon's world to a generalized relativity. It is true that each trajectory is defined not in itself but in relation to any and all of the other trajectories in the novel. Yet the entire constellation, which is imaginary but congruent with flight paths that were actually taken by V-2 rockets, suggests that multiple readings and interpretations can in fact converge on a vanished historical reality. Plurality in Pynchon, and a welter of simulations in and of the so-called natural world, do not in themselves preclude a naturalistic construction of reality.

Through the excessive proliferation of multiple textual realities, Pynchon would avoid imprisoning himself within his own structuring metaphors, in a sense deconstructing his own text in advance. But if in doing so he also resists suggesting a narrative source in some determinate ur-meaning (which no writer is likely ever to recover), he is careful at least to locate each trajectory in its particular historical circumstance. A certain facticity, the historical specificity with which he constructs his postmodern "meteors of style," prevents them from merely rehearsing technology's simulation culture. As readers we might benefit from a similar resistance to interpretive strategies that either totalize Pynchon's textual reality or reduce it to fragments. By remaining attentive to the interplay of large structuring metaphors with the actual working lives of German engineers—from social and psychic managers such as Dominus Blicero and Gerhardt von Göll to the more prosaic rocket technicians Närrisch, Achtfaden, Fahringer, Mondaugen, and especially Pökler—we might begin to move away from the large apocalyptic or deconstructive readings that would equate Pynchon's meaning with the all too evident determinative mechanics of his technique, toward a more localized view of both literary and engineering technologies. In what follows I examine the way Pynchon embeds the engineers' psychological experience within the material forms of their profession and thus particularizes the subject of his social and political critique.

"Strung into the Appollonian Dream": The Peenemünde Engineers

More dramatically than any other object or symbol in *Gravity's Rainbow,* the rocket 00000 embodies the immortal dreams and aspirations of an entire generation of German technocrats, engineers, and rocket scientists. At the same time, it is equally expressive of the personal dream of one man, the German World War II SS officer Dominus Blicero. It is Blicero, "one of several gray eminences around the rocket field" at Peenemünde, who enlists some half dozen engineers of varying expertise to work a modification into the quintuple zero (401). None of the engineers selected for this project has any idea what the rocket is to be used for, that it is being made to hold about its longitudinal axis a human body of a certain size and weight, who will be carried in flight to his death. The rocket is in fact designed for Gottfried, a young Wehrmacht conscript who accepts Blicero as lover, father figure, and imperial leader, and who is literally enamored with the rockets and with the thrill of war they represent. Wrapped absolutely in the Imipolex shroud within the rocket—a perfect fit which gives him a pathetic sense of safety—Gottfried is literally "Strung Into the Apollonian Dream" at the novel's end (754). But Gottfried's fate is only the most extreme and explicit version of a trajectory that is traveled by each of the engineers—Pökler, Närrisch, Achtfaden, and the rest—who work for Blicero. Taken together, they instance the contradictory urge toward personal security and collective extinction that, in Pynchon's view, fuels the Nazi rocket state.

Having appeared in Pynchon's first novel, *V.*, as army Lieutenant Weissmann, Blicero is a surpassing study in colonialist and fascist psychology, a study that begins, for Pynchon, in southwest Africa in 1904, spans both world wars, and looks ahead to the postwar era. In *V.* Blicero is the decadent, mediocre, and technically illiterate junior officer who knows only vaguely that one day engineers will be helpful in advancing the Nazi cause. As he tells the engineer Kurt Mondaugen, "Specialized and limited as you are, you fellows will be valuable" (242). In *Gravity's Rainbow* he reappears as "a brand-new military type, part salesman, part scientist," dedicated to technology as the means of installing Germany in a position of global dominance, and at the same time using the rocket to realize a personal dream of immortality and absolute power (401).[12]

Yet in Pynchon's surprisingly sympathetic portrait, most of which concerns Blicero's dying days in Holland after the war, we are often made to regard this character less with fear or disgust than with a certain nostalgia for a power that, in late 1945, was already passing. Blicero's story, told in Rilkean tones and in a prose that luxuriates in the details of a decadent sexuality, is no less an "Elegy for the World of the Fathers" than the more explicit novelistic deconstructions of a younger writer such as Kathy Acker. Pynchon unambiguously—and somewhat sentimentally—denounces Weissmann/Blicero's participation in the literal enslavement of the Herero people in the Sudwestafrica sections of *V.* In *Gravity's Rainbow* he is concerned with a conception of power that is less familiar, less personal, more abstract, and conceivably more long-lasting than anything Blicero had been able to sustain in Africa or in Germany. Indeed, against those forms of cultural repression that the modern Herero have individually internalized, Blicero's style of frank paternal dominance comes to seem old-fashioned and even quaint. Whereas once (in colonial Africa in 1904) the Herero had been decimated by the Germans, there is now a faction within the expatriate Herero community in Germany known as "The Empty Ones," whose purpose is to "finish the extermination" themselves (317), to bring an entire race to a "Final Zero" and thus free the tribe from its growing subservience to Christian cycles of death and rebirth.

Blicero's sexuality enacts further approaches to the Zero: his elaborate, highly organized sex acts are designed, by those who engage in them, specifically to be "some formal, rationalized version of what, outside, proceeds without form or decent limit day and night" (96). (Unlike others in the Nazi command structure who deplore "degenerate" art, Blicero allows his Dutch slaves all kinds of creative freedom.)

12. Blicero's career closely parallels shifting attitudes toward advanced technology within the Nazi command structure, whose leaders initially failed to appreciate the potential strategic value of the V-2 program but came around later to give the rocket the highest military priority. The difficulties the rocket program faced in winning Hitler's approval are described by Walter Dornberger in chapters 6–12 of *V-2*.

The sex, however, is joyless—an endless, unregenerative repetition for which Katje, "a woman with some background in mathematics" (97) whom Blicero keeps with Gottfried in a casual captivity, finds yet another technical metaphor: "She thinks of a mathematical function that will expand for her bloom-like into a power series *with no general term*, endlessly, darkly, though never completely by surprise" (94). In mathematics a power series is an equation in which some quantity, represented by the variable x, is raised to successively higher powers (e.g., the simplest power series is $1 + x + x^2 + x^3$ and so on). Depending on the value of x, successive terms in the series will either add up to a known sum or go on expanding infinitely toward some known but unreachable value. It is to this second, unbounded or "divergent" series that Katje alludes. A wholly abstract numerical manipulation, the power series is a metaphor for sexual and political actions whose tendency is to replace human beings with inanimate objects and abstract theories, a metaphor, that is, for all those floating constructions and indeterminate sign systems that proliferate in the absence of any sharable social reality.[13]

In ways both actual and symbolic, Pynchon's characters are thus made subject to the same impersonal forces and inanimate objects that mathematics describes and engineering uses as its raw material. "Interpersonal power" in such a world, as in Acker's postcolonial *Empire of the Senseless*, "means corporate power. The multinationals along with their computers have changed and are changing reality. Viewed as organisms, they've attained immortality via biochips. Etc. Who needs slaves anymore?" (Acker 83).[14] For Pynchon as for Acker, the new

13. The mathematical metaphor is anticipated in *V.* when the composer Porcepic works out a chart of "permutations and combinations" listing possible sex roles that the lady V. and Melanie de L'Heuremaudit might adopt and exchange. Another character, who attempts to link sexual decadence to political beliefs that were current in the émigré socialist community in Paris in 1913, is accused of viewing people "as if they were point-clusters or curves on a graph" (see *V.* 380ff.). However we may feel about Pynchon's own tendency to apply abstract mathematical and engineering terms to the construction of character in *Gravity's Rainbow*, these early instances show that he was keenly aware of the social and political implications of what he was doing.

14. A practiced appropriator of the texts of any culture, popular or otherwise, Acker is here lifting a passage from William Gibson's cyberpunk novel *Neuromancer*: "Power, in Case's world, meant corporate power. The zaibatsus, the multinationals that shaped the course of human history, had transcended old barriers. Viewed as organisms, they had attained a kind of immortality. You couldn't kill a zaibatsu by assassinating a dozen key executives; there were others waiting to step up the ladder, assume the vacated position, access the vast banks of corporate memory" (203). Gibson acknowledges a general debt to *Gravity's Rainbow* and in this passage echoes

technologies, whether productive of transcendence or of "despair and nihilism," are larger than any one leader can control. For this reason we might better grasp Pynchon's conception of power by following the lives and individual circumstances of the engineers who help realize Blicero's personal vision, and who carry it ultimately into the corporate culture of postwar America.

Pynchon takes up the story of the rocket technicians Närrisch and Achtfaden at the historical moment in the summer of 1945 when the powers of technological administration and expertise were being transferred, in no very organized manner, from the wartime "Oven State" dominated by Blicero to the postwar "Rocket State" that would culminate in the moon landing of 1969 (the terms are Dale Carter's in *The Final Frontier*). For Närrisch looking on, it is hard to imagine that the work his group did under Blicero could have resulted in so much stir. "Did the S-Gerät program at Nordhausen in its time ever hint that so many individuals, nations, firms, communities of interest would come after the fact?" (517). As long as the war lasted, he was part of a "roughshod elite," "flattered then at being chosen to work on the modification to the guidance, minor as it was" (518, 517). Now he seeks a more glamorous line of postwar work in film, a career move that emphasizes the larger shift in political power away from technologies of production (and mass destruction) toward propaganda and its supporting technologies of *re*production. Having rejected Allied contracts that would have sent him "east with the [Russian] Institute Rabe, or west [with von Braun and other top rocket scientists] to America and $6 a day" (516) he has decided instead to try the gangster life with the megalomaniac filmmaker and blackmarket operator Gerhardt von Göll, a.k.a. "the Springer."

Old habits die hard, however, and Närrisch serves von Göll with the same unquestioning devotion and submissiveness he gave to the Reich: "It wasn't ever necessary to see around the entire Plan . . . really that's asking too much of anyone . . . not true? This S-Gerät strategy he's going out of his way to die for tonight, what does he know of the Springer's *full* intentions in the affair?" (516). With the help of Slothrop, Närrisch has successfully rescued von Göll from his Soviet interrogators Tchitcherine and Zhdaev, and now he is standing cover for the escaped filmmaker and expecting any minute to be over-

Pynchon's argument (through the devil's advocate Father Rapier) cited at the beginning of this chapter. This line of appropriation, from Pynchon to Gibson to Acker, will be discussed further in the epilogue on cyberpunk fiction.

run by advancing sentries. His death, by all indications, is a certain thing, and Pynchon signals its approach with an engineering metaphor taken from Närrisch's home field of guidance. The rocket's automatic steering was controlled by an electronic integrating device known as a Wien bridge, and Närrisch, whose expertise in integrating circuits is celebrated throughout the occupied zone (see *Gravity's Rainbow* 527), finds himself placed under an analogous control:

> The idea was always to carry along a fixed quantity, A. Sometimes you'd use a Wien bridge, tuned to a certain frequency A_t, whistling, heavy with omen, inside the electric corridors . . . while outside, according to the tradition in these matters, somewhere a quantity B would be gathering, building, as the Rocket gathered speed. So, up till assigned Brennschluss velocity, "v^*_1," electric-shocked as any rat into following this very narrow mazeway of clear space—yes, radio signals from the ground would enter the Rocket body, and by reflex—literally by electric signal traveling a reflex arc—the control surfaces twitch, to steer you back on course the instant you begin to wander off. (517)

Närrisch's dream of flight, integrated as it is into the rocket's own device of in-flight guidance, may stand with Gottfried's literal integration into the rocket as the nearest approach in *Gravity's Rainbow* to an image of total technological control over the self (though, in the privacy of the engineer's mind, *he* is the one who exerts control over the rocket, setting upon its brute mass and "fettering" it to his own electronic and mechanical designs [518]). The dream sequence closes only moments before Närrisch is overtaken by the Russians, and he has every expectation that death is imminent. But—and I think this is very important—surprisingly and against all the odds, Närrisch does not die. A few chapters later we learn that Enzian's "Schwarzkommando have got to Achtfaden, but Tchitcherine has been to Närrisch" (563). The engineer's fantasy of kindly Soviet interrogation has evidently come true, and the information he provides under hypnosis has helped fill in the Soviet agent's mental dossier on the rocket.

In itself Närrisch's escape does not have to signify a personal exemption from technological determinism. On the contrary, his preservation may have been part of von Göll's plan—whatever that may be.[15] And there is always the possibility that he was just lucky. Pyn-

15. The Russians, von Göll tells Slothrop later, "weren't supposed to" shoot Närrisch; all the connections were in place if he wanted to escape (527).

chon rarely allows his characters to achieve freedom and individual autonomy through an act of courage alone—certainly not when the character in question is an engineer in the service of the state. The best his characters might achieve through an active decision is the kind of "meta-solution" that Katje effects by secretly leaving Blicero, though in this case she has merely transferred her allegiance from one power to another (102). If Närrisch escapes the destruction that his personal trajectory seems to ensure, it is through no device of his own, but is instead the result of some unforeseen gap or incompleteness within the structure of the system he serves, some failure in the totalizing dream.

This does not mean that Pynchon releases the individual from ethical and political responsibility, or that, like Heidegger in "The Question Concerning Technology," he regards modern technology as "no mere human doing" (300). Technology, for Heidegger, is essentially man's way of giving nature marching orders, though Heidegger's point is that those who drive technology might themselves readily be gathered into the technological order. In *Gravity's Rainbow* the situation of people being absorbed into their own technologies is everywhere in evidence, and the more technical the passage, the more clearly Pynchon reveals how those who would control the means of technological production are in turn controlled by them. Yet while recognizing that, in Heidegger's phrase, we can become so "enframed" by technological devices that we cease even to be aware of them, Pynchon does not absolve humans from responsibility for technology. Writing in the late sixties and early seventies, he is aware of the fascist uses to which Heidegger's philosophy might be put (and the uses to which it *was* put), and he would avoid them by keeping practical, individual actions clearly in view. As the rather heroic, independent-minded technicians among the younger Schwarzkommando argue, it may be "all very well to talk about having a monster by the tail, but do you think we'd've had the Rocket if someone, some specific somebody with a name and a penis hadn't *wanted* to chuck a ton of Amatol 300 miles and blow up a block full of civilians? Go ahead, capitalize the T on technology, deify it if it'll make you feel less responsible—but it puts you in with the neutered, brother" (521).

That "specific somebody" could have been Hitler, on whose sole approval the rocket's priority funding depended, or, within Pynchon's fiction, it might also refer to Blicero. But in assigning responsibility to persons within the Nazi rocket state, Pynchon tends not to dwell on charismatic leaders, whether historical or imagined. The full force of his critique is trained instead on the individual civilians and engineers who served the state, people like Franz Pökler who spent years at Peenemünde "[drawing] marks on paper" without questioning the facility's use of slave labor, and who then moved to the rocket works at Nordhausen without seeking to find out what went on in the neighboring extermination camp, Dora. He "had the data, yes, but did not know, with senses or heart . . . ," he was so locked into his engineering work (432). When Pökler finally does enter the Dora camp and for the first time sees the corpses and near-corpses of prisoners piled one atop the other like so many rockets on his side of the wall, Pynchon leaves us with an unforgettable image of the human cost of Nazi technology.

In the case of the aerodynamics engineer Horst Achtfaden, the question of personal responsibility is treated less passionately but with no less insight into the psychology of wartime rocket engineers. Like Pökler and Närrisch, Achtfaden carries out his own specialized assignments without considering their ethical dimensions and without trying to see around the "entire Plan" (516). Under hypnosis the specialist in first-stage aerodynamics shrugs off any implication that *he* was responsible for the rocket, telling his interrogators among the Schwarzkommando that "he only worked with it up to the point where the air was too thin to make a difference. What it did after that was none of his responsibility. Ask Weichensteller, ask Flaum, and Fibel—they were the re-entry people. Ask the guidance section, they pointed it where it was going . . ." (453). To which the Schwartzkommando respond: "Do you find it a little schizoid [. . .] breaking a flight profile up into segments of responsibility? It was half bullet, half arrow. *It* demanded this, we didn't. [. . .] You are either alone absolutely, alone with your own death, or you take part in the larger enterprise, and you share in the deaths of others. Are we not all one?" (453–54). This is not *exclusively* an expression of a holistic Zen vision, although the voice Achtfaden hears during his hallucination has in fact shifted from the hip technocratic Schwartzkommando to that of the pacifist engineer Fahringer, a convert to Hinduism. The expression of community is equally a recognition that, in a technological world, connections exist that go beyond bureaucratic "segments of responsibility" to form a complex field in which individual action is embedded ineluctably.

In an irreverent parody of specialization, Pynchon has Achtfaden hallucinate the entire passage while aboard that "triumph of the German mania for subdividing," the Nazi "Toiletship" *Rücksichtslos* (the "Reckless"—literally, the "Hindsightless," unable to see its trail of waste [448]). But the parody—one of the strangest passages in *Gravity's*

Rainbow—goes beyond specialization to "satirize the dream about which [Pynchon] shows the greatest ambivalence, that of transcendence achieved through technology" (Tölölyan 145).[16] When the German corporate system breaks down at war's end and Achtfaden is taken by the Schwarzkommando for interrogation (Pynchon's parody of the "de-Nazification" briefings that the Allies put the historical rocket engineers through), he is as uncomprehending of his role in the historical process as Närrisch would be in the face of the Allied occupation of Peenemünde: "What do they want?" Achtfaden wonders. "Why are they occupying a derelict in the middle of the Kiel Canal? Why don't the British *do* something about this?" (451). Pynchon's narrator responds with strange words of comfort: "Look at it this way, Achtfaden. This Toiletship here's a wind tunnel's all it is. If tensor analysis is good enough for turbulence, it ought to be good enough for history. There ought to be nodes, critical points . . . there ought to be superderivatives of the crowded and insatiate flow that can be set equal to zero and these critical points found . . ." (451).

I am not sure that I can do tensor analysis, but then I am not as well versed in aerodynamics as Pynchon evidently had become when he worked up this passage. Pynchon is obscure less often than inattentive readers like to think, but there are times when his technical metaphors refuse to give up their mysteries. Here, in a typical contextual shift, Pynchon wrests the engineer's special language from its proper field of reference and uses it to gather together incidents, data, and perceived coincidences that are important to his own fictive structures and concerns. Like the mythical "self-coefficient" that would express the totality of a character's identity in a single dimensionless number, such historical "nodes" or "critical points" in Pynchon refer the intractable flow of history to the events that occur in a single year. The year 1944–45 (the narrative present of *Gravity's Rainbow*) is one such annus mirabilis, and constitutes the lens through which Pynchon views postwar history. It is to 1904 that Pynchon's narrator refers in the present instance, though the incidents chosen are so obviously arbitrary as to parody his own attempt at historical reduction:

16. The *Rücksichtslos* passage also allowed Pynchon to draw on two areas in which he had direct personal experience: naval operations and corporate engineering. Thomas Moore goes so far as to suggest (without biographical substantiation) that Steve and Charles, "the General Electric apparatchiks who pull an inspection tour of the Nazi Toiletship, [might] originate in Pynchon's days of working for Boeing Aircraft" (174).

> 1904 was when Admiral Rozhdestvenski sailed his fleet halfway around the world to relieve Port Arthur, which put your present captor Enzian on the planet, it was the year the Germans all but wiped out the Hereros, which gave Enzian some peculiar ideas about survival, it was the year the American Food and Drug people took the cocaine out of Coca-Cola, which gave us an alcoholic and death-oriented generation of Yanks ideally equipped to fight WW II, and it was the year Ludwig Prandtl proposed the boundary layer, which really got aerodynamics into business and put you right here, right now. 1904, Achtfaden. Ha, ha! (452)

Prandtl's proposal—whose date must have inspired this breezy historical compilation—is another tour de force of scientific reduction, which allowed the design engineer to restrict the investigation of aerodynamic quantities to the forces present over a thin sheet of air—the "boundary layer"—near the airplane's or rocket's surface.[17] The engineer knows he is ignoring much of the complexity of real airflow, but what matters most is that the simplification *works*. As Pynchon's narrator comments, innovations such as the boundary layer and dimensionless coefficients "[allow] you to use models, arrange an airflow to measure what you're interested in, and scale the wind-tunnel results all the way up to reality, without running into too many unknowns" (453). The engineer's calculations and designs do not describe reality, they model it, and so enable him to create objects and methods that participate in the physical world. To the extent that Pynchon himself has adapted engineering technology to his own fictive purposes, his techniques of personal and historical reduction also participate in the reality they model.

Pynchon's argument is not, therefore, with the engineer's use of reductive *techniques;* he is only too willing to exploit such techniques himself and to construct his fictions in accordance with the engineer's aesthetic. Where Pynchon as a novelist differs from the "moderate" engineers he depicts is in his willing embrace of linguistic excess and consequent refusal to allow his world to be *circumscribed* by technique. He has Närrisch escape the implications necessitated by his own techniques of guidance and control, and he also lets Achtfaden's reductive efforts "breed like mosquitoes" among the Toiletship's proliferating

17. Weisenburger (210) cites this explanation of the boundary layer from von Kármán (61).

waste (453). The dreamlike excursions that accompany each depiction are designed to force the engineer out of his specialized perceptual field and to make him face the social, psychological, and, in Achtfaden's case, historical, context of his work. Without condescending to the engineer, Pynchon is constantly implying important parallels with the artist's own attempt to comprehend reality, whether it be through the act of writing or through the material techniques of architecture, music, and film. Indeed, the proliferation of styles and multiple perspectives on rocket technology in *Gravity's Rainbow* are probably less attributable to some categorizing "encyclopedic" impulse (a term too often favored by readers seeking to claim Pynchon for a recognizably *literary* tradition at any cost) than it is consistent with the endless simulations that an engineer runs in attempting to approximate (and even to produce) the real forces and operations of nature and the machine.

Such simulations are not, like the cultural simulacra catalogued by Jean Baudrillard and other ideologues of postmodernism, expressions of a culture without inherence or persistence, in which anything can be anything else, psychic "identity is untenable," and death alone escapes the game of signification (*Selected Writings* 123). In Baudrillard's world, founded on theories of capital and cybernetic processes not unlike the processes Pynchon ascribes to the German economy of the 1920s, all goes centrifugally, schizophrenically wild by a process of iteration. Pynchon's simulations, by contrast, are acts of the imagination, rooted in engineering operations and concerned with latent meanings in our bodies and in the machines we build. Instead of diverging, value in Pynchon seems to converge paranoically, although value is never "possessed" by any one consciousness. What inheres is the purposefulness of the engineer who scrapes together selected portions of this theory and that and so materializes historical process. It is through constant variation on these engineering operations that technology itself becomes, in Pynchon's writing, not the traditional enemy of identity but a means of integrating the psyche.

4

Technology and Identity in the Pökler Story, or The Uses of Uncertainty

Through the capsule biographies of the engineers in *Gravity's Rainbow*, we have seen how technology can create separate unitary identities that are all too readily gathered into a totalitarian order. Pynchon is especially good at depicting that peculiar form of engineering elitism, "the self-chosen isolation of the technological mind," which Joachim Fest identities as a "[key] to its total readiness to serve" (199). But we have yet to consider Pynchon's characters in a social context, as each one relates, individually, to other people. By isolating figures such as Närrisch and Achtfaden and conceptualizing them in terms of their special disciplines, Pynchon demonstrates technology's potential for fragmenting not only the larger field of contemporary power and knowledge but also the identity of each person within the field. To get beyond fragmentation and elitist technology and fully comprehend Pynchon's attempt to integrate the psyche, we must now account for his representation of intersubjectivity, the self's perception of its relatedness to others.

Here we must look more closely at the two central metaphors from modern physics mentioned briefly at the start of the last chapter, since these metaphors have often been brought to bear (notably by N. Katherine Hayles and Susan Strehle) on discussions of identity and intersubjectivity in many texts besides *Gravity's Rainbow*. Nils Bohr's principle of complementarity, which holds that light can be understood to be either a wave or a particle, depending on the language used to describe it, reminds us that any self-identical thing, when seen from different points of view, can produce entirely different explanations, and the explanations can result in entirely different ways of viewing the world. No less significant for literary studies has been the uncertainty principle of Werner Heisenberg, which says that you cannot measure both the position and the momentum of a subatomic particle at the same time. As models of contemporary thought, both principles can teach us to live with contradictions in our everyday lives and to accept limitations, at the most basic level of matter and light, on what we can say and know about the world. In their largest metaphysical implications, the various scientific models carry subjectivity into the physical universe; they connect events with one another and with the mind of a human observer and so encourage a modern "field view" of reality, which Hayles has identified as "the most important conceptual revolution since Copernicus argued that the earth was not the center of the universe" (*Cosmic Web* 9).

This said, I would nonetheless caution against attempting a too literal translation of scientific terms into novelistic ones. Indeterminacy in literature is not often the same thing as scientific indeterminacy; still less should we expect any writer, especially one so famously subversive as Pynchon, to embrace the field view unreservedly.[1] Academic commentators sometimes forget what a novelist such as Joseph McElroy takes for granted, in "Fiction a Field of Growth," that "the daily *work* of science looks less like trope and turning point and dream and divining than like observation and description. Which in turn reduce the idiom of [scientific] meaning, if not its scope" (1). Heisenberg said as much himself when he wrote, in *Physics and Philosophy*, that "existing scientific concepts cover always only a very limited part of reality, and the other part that has not yet been understood is infinite" (201). The most Heisenberg can say (and it is indeed a great deal) about his work's metaphysical implications is that "modern physics has perhaps opened the door to a wider outlook on the relation between the human mind and reality" (202), but such an expanded outlook is irreducible to "the closed frame of scientific thought" (201).

Hence, McElroy is being neither modest nor overly scrupulous when he insists on the limited idiom of science. Rather, his ambition and his involvement in the field view become, like Pynchon's, that much greater. For it is only by remaining outside the scientific frame that

1. Thus, according to Alan J. Friedman, Pynchon finds the new physics just as "paranoid" as the older, Newtonian model of causal thinking. Both views are embodied in the text as "equal and opposite" madnesses (71).

the novelist can push beyond specialized metaphors and accidental similarities of nomenclature, toward a speculative *use* of science that would be applicable to the whole range of human experience. From this outside perspective, McEloy implies, a novelist might even speculate about the psychological origins of uncertainty itself:

> Who is to say how a scientist thinks? . . . Once upon a time, did Heisenberg have a feeling about our limits in observantly knowing many things—people for instance—acting on each other at a distance, say? Did he have intuitions about how we appear to others because of where they are in time—and was *this* one way he came to his insight about particles he could not see—so he had to rely on what Einstein called "free creations of the human mind"? ("Fiction" 1, 31)

Here McElroy uses physics not as a metaphor, but as an indication of a general *habit of mind* that the writer and the physicist might share, whether or not they embrace either the specialist details or the grander ideas of modern physics. To be sure, a novelist might welcome the introduction of subjectivity into the physical, no less than the social, world, and modern science nicely reinforces literary modernist ways of knowing that have been notoriously abstract, nonintuitive, self-reflexive, and glorying in internal ambiguity and contradiction.

In writing this way, however, the novelist is not necessarily using modern physics to deny the existence of a referential reality, as some of the more programmatic academic postmodernists have claimed. For these critics modern science itself can too easily become the sole referent, and often, even among studies that celebrate Pynchon's destabilizing structures and techniques, conclusions about textual indeterminacy are reached too soon, before the diversity of Pynchon's technical material is adequately covered. Such generalizations, whether they are put forward in the interest of a recuperative-traditionalist or a deconstructive reading, can often bury more detail than they uncover; and as ideas such as indeterminacy and complementarity gain literary currency, the scientific epistemology itself becomes the totalizing system that threatens to limit the kinds of meaning we are likely to find in Pynchon.[2]

2. Gregory Comnes zeroes in on an "absolutist" critical tendency (in the name of greater openness) in a cogent review of Susan Strehle's *Fiction in the Quantum Universe*. Strehle attempts to ground Pynchon's novel, along with narratives by Coover, Gaddis, Barth, Atwood, and Barthelme, in an "actualism" whose referential base

Like all theories that set limits on what we can know, the theories of modern science promote an ironic cast of mind whose imaginative results are various, to say the least, and as likely as not to resist or misappropriate the scientific field. Thus, McElroy can discover conflicting imaginative uses of uncertainty not only in his own and in Pynchon's essays and novels but also in Nicholas Mosley's very different novel *Hopeful Monsters,* which is built on "analogies between what happens to two distant particles and two people," lovers who, once together, remain connected throughout a history of political separations ("Fiction" 31).

Other uses of uncertainty are less explicit and less easily romanticized. A character in Don DeLillo's *Mao II,* for example, speaks of her fascination with the idea of two people walking toward each other from opposite ends of the Great Wall of China, supposedly the only human-made structure that can be seen from outer space. This character knows nothing of the new physics, but she can readily understand the gesture as a story of psychic continuity and "reverence for the planet, of trying to understand how we belong to the planet in a new way." For most of their journey the hikers cannot, of course, see each other or be seen together, but DeLillo's character finds it strange that she can construct "an aerial view so naturally" (70). Such stories, like the field view in modern physics, make us feel the presence of nonobservable and more than metaphorical interactions among people separated from one another in space and time. These stories deepen our sense of human interconnectedness and help sustain a belief that our collective cre-

---

is the quantum view of reality. But this "actualism," Comnes argues, "filters the experience of literature through a concept that is itself privileged, an absolutism emphasizing recurrent narrative structures at the expense of story lines that are markedly different" (216). One notices this absolutist use of uncertainty in many critical studies of Pynchon. David Porush, for example, writes that the only totalizing system in Pynchon "is the absence of a totalizing system" (115), and Molly Hite draws a similar conclusion from Goedel's theorem, which codifies the necessary incompleteness of all totalizing conceptual systems (*Ideas of Order* 124; Hite also locates a number of nontechnical analogues for Pynchon's untotalizable totality, the centripetal trope of the "absent center"). Thomas Moore's metaphor for the text of *Gravity's Rainbow* is the "open system" of thermodynamics (5), and David Seed for his part suggests we take the repeated metaphor of "the lattice and the mosaic" as a model for the book's assembly (158). N. Katherine Hayles, building on the work of Robert Nadeau and Carolyn Pyuen and her own specialist training in science, uses the field concept in modern physics to describe Pynchon's cognitive structures, concluding that the "unruliness" of a book such as *Gravity's Rainbow* ensures that these cognitive structures cannot be complete or perfect (*Web* 169).

ations are capable of producing a connected whole incomprehensible to any one mind in the collective.

This belief is at the heart of the technological sublime, but it is a faith that should not come too easily. That such connections will be necessarily to the good is not immediately clear. For if postmodernist novelists such as Pynchon, DeLillo, McElroy, and Mosley have taught us anything, it is to distrust "this connection-urge," which David Foster Wallace finds "more fundamental and scary than the humanistic syrup of *Howards End*'s '*Only connect*' " ("Empty Plenum" 225n). DeLillo's hikers—performance artists of no particular nationality or politics—cut a line of abstract selfhood through a crowded and distant nation, whose mass of life can only be visualized, by another character in the novel, through the stereotypes of photographic technology: "People trudging along wide streets, pushing carts or riding bikes, crowd after crowd in the long lens of the camera so they seem even closer together than they really are, totally jampacked" (70). Only against this backdrop, a tooled community that threatens the loss of all identity, does the connection at a distance of two idealists become an aesthetic statement of significant, though ambiguous, contemporary power.

Pynchon, too, applies the totalizing language of technology and "images of the new Uncertainty" to crowds, especially those European populations that were forcibly displaced during the Second World War and its aftermath (*Gravity's Rainbow* 303). Indeed, the whole impersonal "bureaucracy of mass absence" is so well suited to the field view of reality that we can hardly accept that view as an unambiguous model of human connection (303). Pynchon's narrator might, for example, monitor the flow of people through the doorway of a London train station: "This wordless ratcheting queue . . . thousands going away . . . only the stray freak particle, by accident, drifting against the major flow . . ." (51). Because this single exception refers to an isolated child who escapes one of the war's deterministic deathward schemes, the metaphor from physics *might* be interpreted hopefully: Pointsman, who has come to the station to search among the displaced nationalities for a young girl to be an experimental subject and possible sexual object, will be disappointed—this time. (Whether or not Pointsman is at the station in a dream or in actuality is unclear.) Elsewhere, Pynchon's use of the scientific metaphor is more ambiguous. A "great frontierless streaming" of human masses across shifting national borders figures the end of balance-of-power geopolitics as well as an end to all romantic and bourgeois notions of the self; yet it is unclear how these people will survive, let alone replace the community they once had (549).

Pynchon's ambivalence about the self's dissolution into crowded social and political fields can be felt in the very texture of his prose, which, as we are about to see, resists romanticization and the powerful seductions of an achieved private style even as it invokes romantic emotions of loss, continuity, and immaterial human connection. These passages depicting the massing of human populations should serve to remind us that the interconnections and reflexive dynamisms of modern physics do not need to be *animate* at all—even if they can or must, in theory, include the mind and body of a human observer. As often as not in Pynchon, modern systems of relationship merely conjure the *ghost* of an earlier romantic notion of continuity and connection. Like engineering technology in the German Reich, the field view of modern science comes to be associated with a mere simulation of life processes, where people relate not with other people but with their likenesses or images:

> . . . barn-swallow souls, fashioned of brown twilight, rise toward the white ceilings . . . they are unique to the Zone, they answer to the new Uncertainty. Ghosts used to be either likenesses of the dead or wraiths of the living. But here in the Zone categories have been blurred badly. The status of the name you miss, love, and search for now has grown ambiguous and remote, but this is even more than the bureaucracy of mass absence—some still live, some have died, but many, many have forgotten which they are. Their likenesses will not serve. Down here are only wrappings left in the light, in the dark: images of the Uncertainty. . . . (303)

In this passage, under the rubric of "the new Uncertainty," the narrator of *Gravity's Rainbow* uses the Heisenbergian motif to suggest a confusion of categories that blurs even the line between living and dead souls. The scene is the Mittelwerke in Nordhausen, which, years before the narrative present, Gerhard Degenkolb had built under a mountain to protect the rocket works from air attack. It was to the Mittelwerke that the Peenemünde engineer Franz Pökler had come "in early '44, as the rocket was going into mass production" (283), to procure raw material for the V-2 and oversee its fast-expanding bureaucracy. But what was then the scene of frenzied activity—the labor that produced the real, historical rocket—has now become a strangely Disneyfied tourist attraction, resembling nothing so much as a buried necropolis:

> . . . once upon a time lathes did screech, playful machinists had shoot-outs with little brass squirt cans of cutting oil . . . knuckles were bloodied against grinding wheels, pores, creases and quicks were stabbed by the fine splinters of steel . . . tubeworks of alloy and glass contracted tinkling in air that felt like the dead of winter, and amber light raced in phalanx among the small neon bulbs. Once, all this did happen. (303)

The factory assembly line is abandoned now—"only the lateness and the absence that fill a great railway shed after the capital has been evacuated"—and the narrative moves, in a passage that recalls the "rush of souls" across an evacuating city at the book's opening, through "lakes of light, portages of darkness," as "entrances to cross-tunnels slip by like tuned pipes with an airflow at their mouths" (303).

Here we have forecast many of the themes, images, and physical simulations that will come to define Franz Pökler's textual reality, in which concrete technological imagery is made to suggest an almost ghostly presence. The narrator's insistence that "all this did happen," and the fairy-tale style so typical of the more esoteric technical passages in *Gravity's Rainbow,* create in the reader a nostalgia for a former, presumably unambiguous reality, now replaced by an absence, a loss of historical reference. Here the simulated ghosts are as real and alive as the connections among people on "this side" who have been separated, in time and space, by the war. In Franz's specific case, "the name you miss, love, and search for" could be either Ilse, his daughter, or Leni, his estranged wife, both of whom had been taken by the SS soon after Franz himself had gone to Peenemünde to work under von Braun, the historical rocket pioneer, and the fictional Lieutenant Weissman (who is later to call himself Dominus Blicero). When Weissmann arranges for Franz to see his daughter, the engineer is never sure whether the girl he meets is really Ilse or some hired model, her "likeness" made to serve.

One of Pynchon's primary images for this uncertainty is the wind at Zwölfkinder, the site of a decaying amusement park, where Franz sits waiting at the start of the Peenemünde chapter. Thomas Schaub has written eloquently of Pynchon's "Orphic Voice"—its elegiac concern with "the connections and continuities of loss and separation" (*Pynchon* 128), its intimations of wholeness in the relations among fragments—and this description applies to the sounds and voices that impinge on Pökler's consciousness. Like the romantic metaphor of the correspondent breeze, the wind here represents a predominant theme of continuity and interchange between subject and object, nature's outer motions and the interior life and emotions.[3] But whereas the central mediating figure for the mind and nature in the greater romantic lyric was the aeolian harp, Pynchon's mediating image is taken from the technology of organ pipes:

> If there is music for this it's windy strings and reed sections standing in bright shirt fronts and black ties all along the beach, a robed organist by the breakwater—itself broken, crusted with tides—whose languets and flues gather and shape the resonant spooks here, the candleflame memories, all trace, particle and wave, of the sixty thousand who passed, already listed for taking, once or twice this way. (398)

Apart from the grander conceptual emphases of modern science, its deepest presence in *Gravity's Rainbow* is to be found in brief passages like this one which would seem to recover a lost historical referent in the plastic and self-referential material of language itself. Initially the sentence looks backward to find its object, "this" referring to the scene at Zwölfkinder, which for Franz is suffused with mystery, its starlight "precarious to him as candles and goodnight cigarettes" (398). The next word, "it's," does not lead to the anticipated description of the sort of music this scene might call to mind, but instead shifts the prose forward into an alternate description of the scene, a renewed vision of its elements. The rusted iron amusement park animals, "their heads jittering with air currents," turn into "windy strings and reed sections"; black snakes on a background of painted sunlight become, with a little imagination, the black ties over the players' "bright shirt fronts"; the breakwater, "itself broken, crusted with tides," at night after ebb tide may become the gathering flue pipe (the ambiguous "whose," like the shifting "it's" before, helps to bring about this transformation).

Pynchon is doing more here than simply introducing music imagery to complete his composition of the scene at Zwölfkinder. The music is less a part of the external landscape than it is created in a process of

3. M. H. Abrams discusses the wind motif in the greater romantic lyric in "The Correspondent Breeze." Elsewhere I have noted Abrams's probable influence on Pynchon when he was an undergraduate at Cornell University ("Pynchon's 'Entropy' " 61).

verbal transformation: "If there is music," if there is an order beyond the initial scene that remains unformed before Franz, then that music is at once called up by and heard in the gathering and shaping of the prose. Independent of Franz's or any single consciousness, one world is transformed into a simultaneous alternative world; the scene and the memories, as if transformed into sound, are given a sense of presence, and the resonant spooks exist literally on the page: "here."

What is unique about this self-reflexive moment in Pynchon (which distinguishes it from other devices of reflexiveness as they are most often used by other modern and postmodern writers) is that the writing is neither wholly mimetic nor self-absorbed and insular. Instead of becoming mired in its own mechanisms, Pynchon's prose gathers and shapes a reality that is external to it. The "influence" of modern science, if one can even speak of something so causal as influence in this passage, has very little to do with metaphorical borrowings or even structural similarities. Rather, as in passages from McElroy's novels that I discuss in the next chapter, the very *processes* of Pynchon's language parallel and reproduce the reality implied by the forms of modern science. Hence it is appropriate, but not essential, that Pynchon again invokes the language of modern physics—"all trace, particle and wave"—to figure human gatherings and dispersals. Such language, with all that it implies for the interdependence of conflicting modes of description, is more than metaphorical, for the passage in which the scientific allusion appears creates a self-referential field that is homologous with modern scientific thought, where "everything," according to Hayles, "is connected to everything else by means of the mediating field, [and] the autonomy assigned to individual events by language is illusory. When the field itself is seen to be inseparable from language, the situation becomes even more complex, for then every statement potentially refers to every other statement, including itself" (*Cosmic Web* 10).

The questions that follow the Zwölfkinder transition reveal the emotional depths of such an image of personal connection over time:

> Did you ever go on holiday to Zwölfkinder? Did you hold your father's hand as you rode the train up from Lübeck, gaze at your knees or at the other children like you braided, ironed, smelling of bleach, boot-wax, caramel? Did small-change jingle in your purse as you swung around the Wheel, did you hide your face in his wool lapels or did you kneel up in the seat, looking over the water trying to see Denmark?

> Were you frightened when the dwarf tried to hug you, was your frock scratchy in the warming afternoon, what did you say, what did you feel when boys ran by snatching each other's caps and too busy for you? (398)

Few readers will recall from some 250 pages back that Leni grew up in Lübeck. More likely we experience Lübeck in this passage as one of several concretizing details that shift the prose away from mounting abstractions and into a vivid north German scene, which Pynchon's narrator is self-consciously trying to make real. This shift, along with the use of the second person, invites the reader to be gathered among the sixty thousand, "the other children like you," and perhaps evokes within the reader's response the perception, so rarely attained by Pynchon's characters, of interlinked human experience.

Franz himself, though he does not see it, is also gathered among the Zwölfkinder revenants. Earlier in the novel a brief parallel passage had described Franz as a child with his parents riding a train not from Lübeck to Zwölfkinder but to the Rhine falls, where he "held on to both their hands, suspended in the cold spray-cloud with Mutti and Papi, barely able to see above to the trees that clung to the fall's brim in a green wet smudge" (160). The connections that the prose only hints at are never perceived by Franz. His own memories remain for him apart from the sixty thousand and from Leni, though he does at times think of her condescendingly as "his own ghost," a version of his former idealistic self (399). Franz's childhood memory was called up in response not to other people but to a static test at the rocket field in Reinickendorf: "These were the kinds of revenants that found Franz, not persons but forms of energy, abstractions . . ." (161).

As metaphor I think we can identify these "forms of energy" with the forces unleashed by modern physics and technology. Surely they are the same forces that Henry Adams had observed in the electronically animated dynamo and in random processes of molecular vibration and nuclear radiation that had threatened, as early as 1905, to break down the predominantly linear, causal, and rationalist constructions of an earlier Newtonian worldview. Like Adams, Pynchon measures all such "mysterious energy" by its attraction on the human mind (*Education* 383), so the breakdown of the Newtonian paradigm may be reflected not only in the physical world and its narrative representations but also in an individual's private failure to construct a coherent personal identity. Hayles seems exactly right to identify the field concept in *Gravity's*

Rainbow "as an unavoidable, and ultimately tragic, double bind" (*Cosmic Web* 11). And the phrase "double bind" is particularly suggestive since it could as easily refer (although Hayles does not pursue the idea) to the psychology and systems theory of Gregory Bateson as to the physical field. Indeed, the structure of identity in *Gravity's Rainbow* might owe as much to Bateson and to his more popularly known student R. D. Laing as it does to Heisenberg. Whatever the conscious source or inspiration of Pynchon's dual imagery, the homologous structure takes us beyond the epiphenomenon of either modern psychology or physics, beyond the contradictory surface details of Pynchon's text, to the sublime structure that permeates all three fields—the scientific, the psychological, and the aesthetic.

Specifically, Franz's divided subjectivity might be better expressed not in the grand terms of metaphysical uncertainty but in terms of "ontological insecurity," a phrase of Laing's that covers various models of psychological fragmentation and dislocation. Franz's unconscious dread of relationship with his daughter, Ilse, can be described as an attempt to prevent "engulfment" of his identity by another person: "A firm sense of one's own autonomous identity," writes Laing in *The Divided Self*, "is required in order that one may be related as one human being to another. Otherwise, any and every relationship threatens the individual with loss of identity. One form this takes can be called engulfment" (45–46). Franz constructs his identity and emotional vacuum so rigidly, however, that he constantly feels threatened by the least perturbation, as if by a wind of energy and abstraction. Such feelings produce the inverse of "engulfment," a second ontological condition that Laing calls, emphatically, "*implosion*" (47). The melodramatic word is particularly appropriate to Franz's situation, since Pynchon, no less than Laing, wants to suggest "the full terror of the experience of the world as liable at any moment to crash in and obliterate all identity, as a gas will rush in and obliterate a vacuum. The individual feels that, like the vacuum, he is empty. But this emptiness is him" (Laing 47).

Pynchon uses just this imagery of a vacuum and a gas to express Franz's emotional hollowness. When, for example, the engineer is confronted with the unsettling possibility of experiencing a selfless love for his daughter, he feels the void inside, and "the vacuum of his life threatened to be broken in one strong inrush of love" (407). Franz maintains his internal vacuum, however, with seals of paranoid suspicion, inventing complications to insulate him from forces that threaten to subsume his isolated self. He imagines that the girl who, after years of separation, has appeared at his quarters in Peenemünde is not Ilse at all, but an Ilse look-alike sent by Weissmann for reasons he will never fully understand. Franz opts to wait out this game to see what variations the SS officer is going to try next, glad in the meantime to have his anger at Weissmann "to preserve him from love he couldn't really risk" (408).

Despite the protective workaday buffer of his engineering craft and language, at the outermost edge of Franz's senses and emotions remains the fear of losing himself entirely to the rocket, as when, during the early test rocket blast at Reinickendorf, there was "no way for the moment of knowing if he was still inside his body" (161). Although he is usually able to suppress the cold knowledge that all of his work and many of his actions will ultimately contribute to human death on a vast scale, Pökler does have a moment when he senses the full implications of his "guiltmaking craft":

> Because something scary was happening. Because once or twice, deep in the ephedrine pre-dawns nodding ja, ja, stimmt, ja, for some design you were carrying not in but *on* your head and could feel bobbing, out past your side-vision, bobbing and balanced almost—he would become aware of a drifting-away . . . some assumption of Pökler into the calculations, drawings, graphs, and even what raw hardware there was . . . each time, soon as it happened, he would panic, and draw back into the redoubt of waking Pökler, heart pounding, hands and feet aching, his breath catching in a small voiced *hunh*—Something was out to get him, something here, among the paper. The fear of extinction named Pökler knew it was the Rocket, beckoning him in. If he also knew that in something like this extinction he could be free of his loneliness and his failure, still he wasn't quite convinced. . . . So he hunted, as a servo valve with a noisy input will, across the Zero, between the two desires, personal identity and impersonal salvation. (405–6)

The alternation between a sense of self embedded in a silent and isolated void and a sense of self lost in an overwhelming ether (not of sound and imaginative warmth—the "Soniferous Aether" [695]—but of energy and abstractions [161]) sets up a fragile tension at the nighttime edge of Franz's perceptions. The engineer's waking dream enacts Laing's psychodrama of engulfment and implosion in the mind's alternation between "personal identity and impersonal salvation." But in this passage we also have a link to the more general psychology of the sublime, for Pynchon's figure of the "servo valve" recalls Kant on the mind's relation with the sublime in nature. In both cases a psychic ambivalence in the face of irresistible power is expressed in the language of mechanism and energy. The mind experiences (in Kant's words in "The Analytic of the Sublime") "a vibration . . . a rapidly alternating repulsion and attraction produced by one and the same Object"—not, in Pökler's case, a natural object, but the rocket that the engineer helps to create (but which he cannot hope to understand in its entirety). The very details of a technological reality, "the calculations, drawings, graphs, and even what raw hardware there was," create in Pökler a sublime psychology, which posits "a wish" (to quote Thomas Weiskel's extrapolation from "The Analytic of the Sublime") "to be inundated and a simultaneous anxiety of annihilation" (105). "Fascination and dread coincide" (104), and the ego is held between opposite attractions—a solipsistic withdrawal into inert mechanism and a desired contact with human beings that cannot be sustained emotionally.

For Pynchon as a writer, this "nighttime edge" is yet another experimental interface between discourse systems, where the combined languages of human psychology and technological mechanism (and the related modes of ontological insecurity and scientific uncertainty) can be made to express more than either language alone could do. But Franz himself only rarely glimpses these edges—"in the ephedrine pre-dawns" when the drug begins to fail and he experiences his loneliness and failure in all their waking clarity. The rocket's promise of transcendence is revealed to him here as indistinguishable from its potential for extinction. But Franz draws back from the realization in fright and lives most of his days using "the gift of Daedalus," the first engineer, to fortify his labyrinthine redoubt. He moves from the role of passive victim to an active agent in the rocket's production. The renewed activity allows him to sustain a sense of a personal identity, but such a displaced and self-defensive identity ultimately leaves him unable to connect with other people, and often in doubt as to their reality.

Franz engineers his emotional life with the same care. His bursts of love for Ilse are renewed at each of the yearly summertime visits the Reich allows. Franz, whose engineer's training alerts him to analogies and correspondent models, is quick to find images that imply "his own cycle of shuttered love." The nearest at hand is the wind tunnel at Peenemünde, where he would stand "listening to the laboring of the pumps as they evacuate the air from the white sphere, five minutes of growing void—then one terrific gasp: 20 seconds of supersonic flow." Franz, by an act of will comparable in its exertion only to the force of the pain it holds back, broadens *his* time base to a year, and so builds "the illusion of a single child" (422). The key thing for him is always to perceive his daughter's identity as an *illusion*, and he is willing to go so far in this as to allow that the child he sees may not be the same child each year, but one of several thousand acceptable "Ilses" available to the Reich. For as long as Franz suspects Ilse's status as a single identity, his periodic bursts of love can be damped down by suspicion to safe, controllable levels. The vacuum of the wind tunnel represents for him the subjective void surrounding and preserving his isolated center of personal identity, and if it is to remain intact, if he is to prevent the implosion that would obliterate all identity, he cannot risk an unrestrained emotional contact with another person. Franz attenuates his perceived reality of a daughter's single identity so that he can preserve the constructed daytime reality of his perceiving self.

The ontological status of Franz's perceptions is made yet more uncertain by another analogy for his relationship with Ilse. A great movie buff, Franz thinks of the persistence of vision that holds between successive frames as they flash on the screen to create for the viewer an illusion of continuous movement. Back in Berlin before the war, he would "[nod] in and out of sleep" as he watched ordinary movies, using not the physical connective of persisting vision but extraordinary powers of cause-and-effect thinking to "connect together the fragments he saw while his eyes were open" (159). By thus threading occasional glimpses of moving pictures, he would construct his own rarefied but continuous story.

The film metaphor expands to include other, less bizarre ways of constructing artificial continuities: the technicians at Peenemünde, for example, use cinetheodolite photographs—pictures of the rocket taken at discrete points along the continuous arc of its movement—to "counterfeit" an image of the rocket's flight. The narrator notes that, using a related analytic technique, which was to result in the invention of calculus, Leibniz had broken up the trajectories of cannonballs in flight, imposing artificial divisions in time to bring the infinitesimal moment into the realm of rational analysis. At one point the narrator asks if this is not "every paranoid's wish? to perfect methods of immobility?" (572), and this characterization is well suited to Franz's dissection of motions, for he employs the technique to help fight his growing fear that the SS officer Weissmann and the rocket are both "out to get him" (406).

The movie analogy, then, as it is used in the Peenemünde section to describe Franz's relation to Ilse and the rocket, is one of Pynchon's devices for calling into question our insistence on framing experience according to traditional mechanistic (i.e., Newtonian) notions about causal connections. It seems reasonable to conclude, with Mark Siegel (82) among others, that, because of their insistent imposition of artificial causal connections onto the flux of experience, Franz and the other paranoid or scientific analysts enter into a compromise with reality. Robert Nadeau adopts this view when he writes that Pökler "as a rationalist is incapable of intuiting that all human activity is interconnected and also that much of that activity is irrationally motivated" (144).

And yet, if Pynchon accuses Pökler and the other Peenemünde engineers of striking a compromise with reality, he makes it difficult for readers to go beyond this conclusion and determine just what, in the context of *Gravity's Rainbow*, this violated reality might be. For if Franz's perception of Ilse as a single identity is elusive, ours is apt to be little more precise should we attempt to construct for her a stable and coherent identity as a character in fiction. We are never told, of course, how valid Franz's suspicions are, but the uncertainty about Ilse's identity is even more deeply ingrained in the narrator's presentation of her. At one point, for example, her identity is commingled with that of Gottfried, Weissman/Blicero's catamite, for just as Ilse once delighted her father with dreams of building a house overlooking the "seas" of the moon, so would Gottfried whisper Blicero to sleep "with stories of us one day living on the Moon" (723). Like Blicero, Franz laments his child's abandonment of the dream.

This connection is not an isolated coincidence but part of a bizarre "system of analogies among characters and events" stemming from film director Gerhardt von Göll's erotic thriller *Alpdrüken* (McHale 79).[4]

4. Brian McHale's discussion in "Modernist Reading, Post-Modern Text" (reprinted as chap. 3 of *Constructing Postmodernism*) is still the most interesting treatment of the subject of Pynchon's analogical correspondences. McHale concludes that Pynchon's proliferating analogies constitute a parody of the modernist device of analogical integration. The parody, however, does not automatically free us into an alternative world of infinite possibility: "No doubt it would be difficult to live in

The star of that movie, Greta Erdmann, conceived a daughter, Bianca, during the filming of a gang rape, and Franz had Erdmann's image in mind when he fathered Ilse on Leni. The narrator suggests that Franz was not the only one so affected ("How many shadow-children would be fathered on Erdmann that night?" [397]); perhaps the film spawned an entire generation of German children.

At one level Ilse and Gottfried are thus linked to each other as versions of Bianca on this side of the screen. But the analogies keep proliferating. The *Alpdrüken* scene, far from being the "real" source underlying the creation of illusive characters, is itself patently illusive because in filming it von Göll, who was experimenting at the time with gnostic symbolism, used a double lighting technique that gave each actor two shadows. (For reasons known only to von Göll, one shadow is for Cain and one for Abel.) At this point Pynchon has sufficiently undermined the usual distinctions between film and reality to halt the profusion of layers and invite us into a world grounded in analogical correspondences, insisting that above all of von Göll's expressionistic imagery the connections between the living shadow-children are real. The children persist on this side of the mediating screen beyond the film's end, "not out of any precious Göllerei, but because the Double Light was always there, outside all film, and that shucking and jiving moviemaker was the only one around at the time who happened to notice it and use it" (429). The double connections between characters that exist "outside all film" are then adopted as a disorienting but consistent mode of reality inside *Gravity's Rainbow*. It is thus possible, when working under such a conception of reality, for the narrator to make a comment like this: "Ilse, fathered on Greta Erdmann's silver and passive image, Bianca, conceived during the filming of the very scene that was in his thoughts as Pökler pumped in the fatal charge of sperm—how could they not be the same child?" (576–77).

The relationship between reader and character in this new mode of reality is in effect the inverse of the relationship between Franz and his daughter: whereas Franz connects multiple year-to-year images of a child with a single name, Ilse, we are led to blend into *the same person* a series of characters with different names. Ilse's stability is for us no less illusive, her identity no more coherent, than it is for Franz. In

---

an ontologically unstable world like that of Pynchon's novel, where things flicker between reality and unreality" (88; see my discussion of the "shadow children" Ilse, Bianca, and Gottfried, which follows).

semiotic terms, the mind's relation to reality breaks down from either an inundation or a deprivation of meaning, analogous to Franz's earlier alternation between identity and an "impersonal salvation" that destroys identity. For Franz the relation to the real collapses in apparent excess on the plane of the unknowable human subject—the signified—and for readers the excess is experienced on the plane of signifiers, the complex of names: Ilse/Gottfried/Bianca. In either case the children's technologically induced, filmic origin liquidates their reality.

This is not a situation peculiar to these particular characters whose interconnections I happen to be tracing. Brian McHale has shown "that almost any character in this novel can be analogically related to almost any other character—to raise for us the demoralizing prospect of free and all but unmanageable analogical patterning" (80). Is this, then, the "real" world of *Gravity's Rainbow* that as readers we are meant to inhabit? Much in Pynchon criticism published since McHale's 1984 essay would suggest that it is, that McHale is naive to be demoralized, that he is a technocrat for preferring manageable books, and that the distinction we have been worrying "between the cinematic and the real" is moot, indeterminable, reducible to only a question of "text" and "more-text" (McHale 83–84, citing McHoul and Wills 49). I believe that McHale rightly rejects such textualist readings as too quickly asserting the necessity of uncertainty—in this case the undecidability between real and filmic identities. Not only do such readings ignore the need to make hard choices even when the ground for choices has disappeared; but on their own terms, postmodern textualist readings forfeit a certain textual *pleasure* in the discovery of uncertainty, which comes from "the need," in McHale's words, "to gauge the evidential value of any given passage, the whole rich hesitation between competing hypotheses" (89). Still more, apart from the epistemological problems of processing the textual reality, there remains the question of the desirability of living and interacting (not to mention *writing* coherently) in such an ontologically unstable world.

The difficulty again is that freedom in such textualist readings is conceived wholly negatively and ironically, and is thus ultimately dependent on the continued existence of the very rationalist construction they would be free of. There is no positive presentation in the novel of anything like an analogical "community" or nonrational system of human relations. If we move, for example, beyond the Pökler story to consider a character in *Gravity's Rainbow* who lives in a world constructed wholly from analogical connections, we come to an ambiguity no less disturbing than Franz's rationalist construction. Miklos Thanatz, who finds himself adrift in a sea of shadows and accident, is not sure "even if [Gottfried and Bianca] aren't two names, different names, for the same child . . ." (671). Thanatz's situation is precisely the inverse of Franz's, for where Franz maintains a suspicion about Ilse that attenuates *his own* love, Thanatz feels himself suspended before Bianca in "perpetuate doubting of *her* love—" (672; emphasis added). He is not isolated, in doubt as to his own self-consistent reality, but he does question the reality of the indistinguishable others: "When mortal faces go by, sure, self-consistent and never seeing me, are *they* real? Are they souls, really? or only attractive sculpture, the sunlit faces of clouds?" (672; emphasis added).

Taken alone, either of these two modes of representation in *Gravity's Rainbow*—causal connection and analogical integration—will ultimately represent characters in perpetually uncertain relations to their world. Thanatz's connections with other people merely lose him amidst the flotsam of an indistinguishable and lost humanity, a condition that is no better than Franz's isolation. And since the novel never explicitly denies or sanctions either mode but simply presents both, this uncertainty extends to the relation between the reader and the text. Just as Franz, in all his ontological insecurity, must choose between "personal identity and impersonal salvation" (406), so the reader is left to choose between a world and its inverse, in which one of two conditions holds: either the mind is overwhelmed by meaning in the world outside consciousness (the signified), or a perceived absence of determinate meaning is filled up with a purely linguistic construction—an excess on the plane of the signifier. The choice is between identity and absence, between a mind that is inundated by external, consensually determined meanings and one that pushes out into a void, replacing it with analogical connections, with metaphor, with any of its own (more or less desperate) fictions of continuity.

Thus, we have moved from the psychological model of ontological indeterminacy to a semiotic model, an alternation between analogical and causal constructions that Weiskel has identified as the two poles of a "metaphorical" and a "metonymical" sublime:

> We may call the mode of the sublime in which the absence of determinate meaning becomes significant the *metaphorical* sublime, since it resolves the breakdown of discourse by substitution. This is, properly, the natural or Kantian sublime, and we might think of it as the herme-

> neutic or "reader's" sublime. . . . The other mode of the sublime may be called *metonymical*. Overwhelmed by meaning, the mind recovers by displacing its excess of signified into a dimension of contiguity which may be spatial or temporal. (28–29)

It is tempting to map the two poles of the sublime onto Pynchon's famous opposition between paranoia and anti-paranoia, conditions in which both "nothing" and "everything" is connected (*Gravity's Rainbow* 434). In the metaphorical mode of narration and/or perception (the "reader's sublime," according to Weiskel, and Pynchon's "anti-paranoia"), "nothing is connected to anything, a condition that not many of us can bear for long" (*Gravity's Rainbow* 434). Slothrop in his moment of antiparanoia, for example, "feels the whole city around him going back roofless, vulnerable, uncentered as he is." His frame of mind is *readerly*, ready to insert itself, its linguistic excess, and its own "pasteboard images" into a world it has emptied of meaning (*Gravity's Rainbow* 434). The readerly mind "resolves the breakdown of discourse by substitution," so that meaning comes to be grounded in a fiction, a simulation, a free play of textuality. The metonymical, "writerly" mode corresponds to Pynchon's "paranoia," in which "everything in the Creation" is connected causally (*Gravity's Rainbow* 703). The paranoid mind that is "overwhelmed by meaning" reacts by "projecting" itself outward so that self and world merge in a single contiguous reality (Weiskel 29). In either case the imagination reads or writes itself into all things, and the entire world threatens to become nothing but language.

Tempting—and all too neat—such mapping surely is. But in following these connections to the point where they touch on a central critical theme of the present book, have I not lapsed into a paranoid critical mode of my own? It might well be objected that I have taken the analogies among psychological, semiotic, and scientific models too far, indeed to the point where every ontologically unstable model begins to look like every other model. The interpretive system that I have erected on Pynchon's oppositions threatens to collapse of its own overdetermined weight, much as the identities of the separate "shadow-children" in *Gravity's Rainbow* collapse into a wholly linguistic system of analogies. Let me then leave off the attempts at categorization and simply note the fact that congruence, the layering and simultaneous functioning of homologous discourses, would appear to be Pynchon's mode in *Gravity's Rainbow,* one that is consistent with the overall mode of uncertainty in physics with which this chapter began.

What, then, does the paradigm from physics finally contribute to Pynchon's poetics? In *Gravity's Rainbow,* it clearly helps to define the texture of language in perpetual difference from that which it describes: light in physics is neither wave nor particle, even though the physicist has no choice but to speak of it as one or the other. So, too, in Pynchon's narrative is duality a condition of *language,* not of its subject. This duality does not, of course, solve everything for Pynchon. Such proliferating ambiguities and dualities are after all the established techniques of literary modernity, by which writers (and their explicators) can propose, simultaneously, any number of conflicting views without taking responsibility for any *one* view, a facile pluralism that ends up sustaining the very political order that Pynchon's radical fiction critiques. Moreover, multiplicity and metaphysical relativity have always served modernist writers and their critical professorate in the traditional battle against the Great Satan of determinist science; one should not claim for Pynchon a major aesthetic or conceptual achievement simply because one source of his indeterminacy is to be found in science itself.

Beyond what is after all a fairly conventional modernist indeterminacy, one might however identify a deeper instability in *Gravity's Rainbow*—one of *tone.* Pynchon's narrator, like the character von Göll, may be the "shucking and jiving" director who notices the dual light and playfully allows it to shadow forth endlessly proliferating identities, suggesting not a reality that has degenerated into apolitical relativism, but the existence of an unnamed, deeply felt reality wholly outside any self-consistent interpretation we might bring to it. Readers might continue to admire such ambivalence as an indication of Pynchon's heroic resistance to determining schemes of *any* sort. Yet the unnamed underlying psychological reality that comes through the interstices of these supposedly determinist networks remains unremittingly nightmarish and ontologically insecure—a poor alternative to the discarded "truths" of a more normative worldview.

This structural and psychological "double bind," though it may be the result of heroic resistance, is also indicative of a continuing inability on Pynchon's part to imagine a reality in which *any* human connection is possible, a problem that I believe to be at one with the difficulty of locating a social and material reality that is not merely simulated. Of all the engineers in *Gravity's Rainbow*, and even of all the novel's main characters, Franz Pökler is the only one who is provided with a family (albeit a broken one), a fully imagined historical context, and a sufficient number of friends and professional acquaintances to take part in potential systems of caring and connection. Not until the publication of *Vineland*, seventeen years after *Gravity's Rainbow*, would Pynchon again involve himself this directly in the relations among scientific, technological, and familial structures which Franz invariably draws back from. Indeed, the personal warmth of *Vineland* makes the comparative absence of sustained human relationships in *Gravity's Rainbow*, familial or otherwise, all too evident.

We might understand the new domesticity of Pynchon's recent fiction, then, as an optimistic alternative to an enclosed, protectively rational self on the one hand and an accommodation to stifling corporate connections on the other. (Biographical explanations are also conceivable: Pynchon is, after all, in late middle age at the time of this writing, and is rumored to be a father.) The ordinary domestic life in *Vineland* would seem to be motivated, from an aesthetic viewpoint, by a desire to move away from modernist indeterminacy and technical abstraction toward a "grounding in human reality" that Pynchon, purporting to speak in his own voice, claims he misses in his earlier work (*Slow Learner* 18). This claim is made in a semiautobiographical essay in which he goes on, with disarming directness, to reject those fictions of his youth which he feels lack an authenticity, "found and taken up, always at a cost, from deeper, more shared levels of the life we all really live" (*Slow Learner* 21). Few readers, I hope, could find such authenticity lacking in *Gravity's Rainbow*, but the "deeper" levels and paradoxical coherences in this novel now appear to have been psychological rather than social, and not resolvable into a worldview or vision that could ever be shared publicly (except via communications media which Pynchon now regards as responsible, largely, for his generation's dissociation from any political reality: "We were onlookers: the parade had gone by and we were already getting everything second hand, consumers of what the media of the time were supplying us" [*Slow Learner* 9]).

Unlike the earlier psychological narratives, Pynchon's recent work (and the aesthetic in the autobiographical and critical essays that anticipates and sheds light on the work) now places technological abstraction and media simulation in opposition to "human reality" (*Slow

Learner 18). As a result, the domestic grounding that Pynchon has been working toward may have been unnecessarily limited. *Vineland* represents a disturbingly reactionary aesthetic, one that has moved away not only from the admittedly facile undergraduate abstractions of the early stories but also from the experimental exuberance and linguistic prowess of *Gravity's Rainbow*. In particular, the public persona of the past few years, the "slow learner," "Luddite," protector of guild "secrets," and couch theoretician of "sloth," seems designed specifically to distance the author, if not from *Gravity's Rainbow* itself (for Pynchon is careful in these essays never to name the novel), then from its academic reception.

Of course, this persona was part of his fiction from the start, dating not only from Benny Profane in *V.* but from Profane's prototype, "Lard-Ass" Levine, in Pynchon's very first published story, "The Small Rain." This figure, who would turn up again in Slothrop and in Zoyd Wheeler in *Vineland*, is the counterpart and ironic corrective to the more flamboyant, technically sophisticated artist figures in Pynchon's work—figures who are often either manipulative (like Blicero), unapologetically elitist (like von Göll), ineffectual (like Roger Mexico), or co-opted (like Pirate Prentice and the rest of the counterforce). But Pynchon may have a deeper rhetorical reason for adopting the former, less aggressive persona for "himself." A sublime that recognizes itself as such is no sublime at all, and in the face of meanings and aspirations that cannot be shared, and perhaps not even admitted to himself, Pynchon is doubtless attracted to the humble persona whose most salient admission of self-worth is an expression of wonder that he was the one who wrote parts of "The Secret Integration," one of the better early stories (*Slow Learner* 22).

The irony that was once so powerful a mode of epistemological and political critique is now put into the service of self-deprecation and internal resistance to the technological culture, not to mention his own appropriation by the culture's academic wing. Pynchon's expression rarely exceeds the lukewarm ironies that are now the stock-in-trade not of high postmodernism but of television, corporate advertising, and other contemporary media. For the reader who has been moved by complexities of form and language beyond the irony and alienation depicted in the great early work, the "return home" to family life and an American setting in *Vineland* can seem imperfectly achieved, the new optimism arbitrary, settled in advance, and sustainable only by an almost willed holding back from darker forces and paranoia that still come obscurely through. To find a narrative that includes less restricted conceptions of both domesticity and abstraction, an un-ironical narrative both social and psychological that exists *within* the abstractions of science and engineering, we must turn to the work of Joseph McElroy.

5

Literature as Technology: Joseph McElroy's *Plus*

Beginning with his first book, *A Smuggler's Bible* (1966), each new novel by Joseph McElroy has been routinely compared to work by Thomas Pynchon, William Gaddis, John Barth, and Robert Coover, but McElroy has yet to be accorded either the readership or the critical attention these more canonical contemporaries command. After the publication of his fourth novel, *Lookout Cartridge* (1974), he was praised by Tony Tanner as an "important writer working with extraordinary energy and imagination right at the very boundaries of contemporary fiction; indeed, he is redrawing some of those boundaries, and at times going beyond them" ("Topography" 252). He would go beyond all boundaries in his next two novels, whose original and uncompromising narratives aim at redefining realism. But except for a theoretically advanced group of science fiction readers and writers, *Plus* (1976) remains underrecognized, and the massive *Women and Men* (1987), which Tom LeClair has called "the most significant American novel published since *Gravity's Rainbow*" and a "model of what the postmodern imagination, informed across spectra of scientific learning and artistic sophistication, can achieve," has not won for McElroy the readership he deserves (*Excess* 132).

His books are certainly generous, and in their persisting optimism and nonsatirical use of science they offer a hopeful alternative to the more paranoid fiction of Pynchon, Coover, and other novelists associated with them. Yet McElroy is no less generous and optimistic in his audience expectations, and he credits readers with such intellectual range and curiosity that even those who admire his work have assumed that he is writing *only* for the future. Among cautiously appreciative peers and reviewers, his work is generally bracketed off as "experimental" or "ahead of its time," a fiction that will be appreciated only after we have reached some distant perspective that will better equip us to understand his scientific determinants. In the meantime, an ambitious and potentially major body of work is left open to charges of "unreadability," so that many of the most engaging postmodern narratives are left unread. Critics thus perpetuate a breach between McElroy and his readers that leaves it to some later generation to decide the work's value, while heroic small presses preserve the texts in college editions and the critical professorate patiently undertakes the task of explication. His books, it is true, continue to stand in high regard, but they are kept alive and studied, like tropical rain forests under glass in Arizona, against the day when we might begin to live at ease within the world's greater ecology.

Few academic readers of Barth's *Letters*, Coover's *Public Burning*, Gaddis's *JR*, McElroy's *Lookout Cartridge*, and Pynchon's *Gravity's Rainbow* have attempted to counter Charles Newman's trenchant attack, in *The Post-Modern Aura*, on these books as "extremely frustrated efforts of extremely talented people" (91). He goes on: "These are perhaps the first works consciously written, not for posterity—but *only* for posterity—a true future fiction for an audience which not only does not exist, but *cannot* exist unless it progresses with the same utopian technical advancement of expertise, the same accelerating value, which informs the verbal dynamic of the novels written for them" (92). What LeClair celebrates as an "art of excess" Newman condemns as an "inflationary" fiction whose demands he links with a contemporary preference for "innovation," "increased growth," and solution by "technique" which must ultimately be seen as utopian, ahistorical, or worse: "an act of ultimate aggression against the contemporary audience" (92).

Writers of McElroy's generation, Newman reminds us, all came of age with "a new Managerial class [whose] tendency to disregard the resistance of traditional phenomena, to overmanage," is grounded in "a belief that technological innovation can free itself from history" (91). This is, moreover, the first literary generation for whom there exists a ready-made audience of professional readers trained in modernist techniques of analysis. The work that such writers are thus encouraged to produce—abstract, specialized, and self-consciously aware of its own modernity—can easily lose touch with felt life and fail to create the "depth and momentum which we associate with traditional narratives. . . . [Failing] to give consistent pleasure, [these novels] are sophisticated precisely because they function very much like the primitive brain, eschewing every familiar sentiment and facility of absorption" (91). "Consciousness," he concludes, "does not progress infinitely any more than profits, productivity, or moral betterment" (92).

Newman is right, of course. But I do not think these novelists are simply trying to compete with technology in the management and development of their own medium. Literary technique is, rather, a way of apprehending *from the inside* the form of contemporary technologies and of aligning consciousness with them. In this respect the condition of the primitive brain may be precisely the sensibility that the ambitious modern novelist aspires to, the most promising tendency of modern technologies being not toward some utopian future or atavistically to the unconstructed past, but toward present and multiple connections that would restructure the presumed holistic relation of the primitive to the world. Indeed, McElroy in *Lookout Cartridge* reminds us that consciousness *can* change, or at least become newly situated with the development of new technologies, when he has his narrator, Cartwright, think of the Giant of Cerne, a 181-foot-long figure of a man cut into the chalk slope of a seaside mountain in England. Flying over the mountain in a plane, Cartwright realizes that the form he sees must have existed before anyone could view it whole: "I tried again to get an over-all view and wondered what the original incisers (Roman Britons or earlier cultists) had been able to see without the moving wand of a plane's aerial height" (152). We may collectively participate in the creation of forms more connected and unified than we can know, and our sense of such connection need not always be paranoid.

Like the conceptual world of modern science, modern narrative often escapes common sense, intuition, rationality, and representation; increasingly it depends on abstraction to order concrete phenomena that cannot be grasped by the human mind and imagination. What distinguishes McElroy's abstraction, however, is a deep attention to the many kinds of language we use in order to *be* abstract. These include much more than the inflationary rhetoric of a bureaucratic society. "Science and technology," McElroy told LeClair in an interview, "offer forms by which we can see some things clearly; their experimental and measuring methods, their patterns larger than life or smaller than sight, beckon us out of ourselves" (LeClair and McCaffery, *Interviews* 238). Far from being self-indulgent or noncommunicative, McElroy uses specialized information and abstract language to render things outside himself, outside the precincts of his own control and literary manipulation.

I do not think there has ever been a more abstract novel than *Plus*. At the same time, few novels pay more precise attention to the experiences that go along with the abstractions of a technological world. *Plus* is the story, quite literally, of a mind in space attempting to reconstitute itself by interacting with the outside world, the "more all around" that the mind can never fully comprehend (12 passim). In an imaginary but frighteningly plausible experiment of the future, a human brain has been surgically removed from the body of a dying engineer, set in orbit, and used as an organic computer to transmit and receive messages from earth. The brain is linked through an elastic bath and life-supporting net of tubing to various plants and food solutions and through a collection of silver electrodes to a solid-state computer and communications system. The electronic linkup enables parts of the brain to transmit and receive "message pulses" on a radio frequency to and from "Ground," where a technical group monitors the effects of sunlight on the biological organism, as in the Skylab experiments of the early seventies (*Plus* 3; McElroy covered these experiments in "The Skylab Cluster"). While this purely functional activity is going on, however, the brain initiates an untransmitted private conversation with itself, which the reader is allowed to overhear through "a highly original use of free indirect discourse" (Brooke-Rose 269). These thoughts make up the narrative, and as they begin to direct the brain's physiological growth (as "mind" is known to affect the physical structure that we call the brain), the organism develops hybrid limbs, strange physical extensions, and powers of sight that penetrate and encompass its world.

McElroy has said that *Plus* is "literally an experiment that comes out surprisingly," but it is not experimental or "literary" in an artificial sense ("Midcourse Corrections" 29). Growth in *Plus* is more than a metaphor, and technology is more than what the book is "about": both material processes, the technological and the linguistic, are congruent with what McElroy calls "the building or the rebuilding of the language capability of this being" ("Midcourse Corrections" 30). The language, which he intended to be literally constructive, *is* the experience of the subject as it builds itself back into systems and structures that it can formulate abstractly but never wholly comprehend. This is quite the reverse of the postmodern frustration of meaning that Newman, among many others, has attacked. As a compositional act *Plus* actually participates in a technological process; its introspective mind-space reflects and reassembles familiar structures in the world. From this perspective the relation of technology to McElroy's fiction can be understood not simply as an experimental mixing of disparate linguistic modes, one technical and the other literary, and not as a metaphorical relation between a scientific tenor and a nonscientific vehicle, but as a relation of similar things: language and technology as collaborative though independent modes of thought, both of which carry and embody power.

Literary and Technological Background

"The Space Program isn't the way to understand *Plus,*" McElroy said in a 1990 interview, properly directing critical attention away from "the hardware" of space technology to the novel's conceptual achievement, and referring to the cultural climate that is created by, and that in turn helps to create, technology ("Midcourse Corrections" 31). *Plus* is not a "researched" novel in the same sense as Tom Wolfe's *The Right Stuff* or Mailer's *Of a Fire on the Moon,* whose authors studied flight manuals, newspaper accounts, and astronaut biographies; nor does McElroy attempt the kind of immediate metaphorical relation of rocketry to human psychology that Pynchon achieves in *Gravity's Rainbow*. Never so programmatically reticent as Pynchon, McElroy is nonetheless reluctant to comment on source processes.[1] And while his work, as we shall see, is to a large extent *about* our desire to create origin stories, the novel's own conception cannot be so easily fixed.

Apart from predictable critical reductions of *Plus* to either a biotechnical catalogue or an empty formalist "experiment,"[2] what necessitates this disclaimer about technological origins is the considerable attention McElroy did in fact pay to the *Apollo* flights and Skylab. In the years leading to *Lookout Cartridge* and *Plus,* he made more than one trip to Cape Kennedy and published a clutch of reviews on the space program

1. In answer to my inquiry about his technological sources for *Plus,* McElroy replied: "While I doubt that ad hoc research ever makes a big difference to the imagining and thinking-through of an original novel, I often find that reading, which is experience too, confirms a fantasy, a thought, a passion, an inkling (cf Wm James), or extends the tentacles and filaments thereof. . . . I don't 'do' science, but sometimes fantasies that make sense to me may be a modest form of experimental thinking" (letter of August 3, 1989).

2. Brooke-Rose (288) dismisses an early paper by Mas'ud Zavarzadeh which calls the language of *Plus* "computer language."

in the *New York Times Book Review*, an article on Skylab in Brooklyn's *Poly Prep Alumni Review*, and several pages on the aesthetics of technology in a 1975 *TriQuarterly* essay, "Neural Neighborhoods and Other Concrete Abstracts." These pieces involved him in much more than fact collecting. At the time they were the closest he had come to asserting a "position," and his choice of journals suggests that in midcareer he may have been publicly associating himself with the space program not precisely to widen his audience but to reach specific audiences in the nation, in his home borough, and in a quarter of the literary academic community that was still devoted to formally ambitious writing and in which his work had the best chance of being appreciated.

The most important of the space essays is "Holding with *Apollo 17*," a brief account of the delayed launch of December 7, 1972. Writing three years after *Of a Fire on the Moon*, when the immediate drama of the first moon launch had passed and the public had become, by all accounts, more or less bored by the space program, McElroy could dispense with what he would later call the "blast-off kitsch metaphysics" of Mailer's "instant Manicheeism" and concentrate on the "central substance of space technology"—on what occupied the astronauts and "the people servicing the rocket. The machines and how they work" ("Neighborhoods" 210). Though a novelist, McElroy claims not to have gone to the launch so that he might write a book that "would capture and crystallize . . . the human side of space or the American dream in what's been happening at Cape Kennedy" ("*Apollo 17*" 27). The conventional journalistic approach presumes a distinction between scientific fact and "human" value that can only perpetuate feelings of alienation. Neither would he indulge Mailer's rhetorical posturing, in which an aggressively metaphorical style is put up *against* technology. Describing the Vehicle Assembly Building, McElroy seems rather at pains to avoid any elaborate or fine writing, as if he distrusted the specifically literary distortions that a metaphorical language might work on our perceptions:

> The VAB: it is where the great parts are "mated." The building's own coordinate parts, its high and low bays, its transfer aisle, its 456-foot-high upward-sliding doors, are designed for the placement and shifting of stages and payloads which are as complete as they can be individually when they come here from Long Island and California, Boeing and Grumman, McDonnell Douglas and North American

> Rockwell ("Where Science Gets Down to Business"). Through triangles upon rectangles of girders and elevator shafts, vistas shadow vistas. Across low-lit complicated drops one sees through into permanent floors like an Erector-Set doll's house so brightly illuminated through the missing fourth wall that two distant men talking and pointing seem on a set. Everywhere measurement and vista conjoin. ("*Apollo 17*" 28)

The passage seems designed specifically to do without any inflationary rhetoric of the self, while the functional metaphors of the stage and Erector-Set serve primarily as aids to visualization. Throughout the essay, NASA terms and corporate phrases are quoted without comment by the author: he forces no additional or too obvious meanings on the engineering term "mated," and he is content simply to cite the public relations tag of North American Rockwell without attempting to challenge the more detailed collaborations between science and a highly militarized industry. This refusal to be distracted by the politics and possible "human" drama in technology's corporate background—the chief concerns of a fast-aging New Journalism—does not mean that McElroy is uninterested in such matters. Relations of power—"power shown being acquired from sources where it had momentum but not clarity"—would be a main theme and governing trope of *Lookout Cartridge,* the novel McElroy was completing when he wrote the *Apollo 17* essay (*Lookout* 77). Here, whatever interest the novelist might have in such issues would wait on an accurate perception of the literal structure before him, its power seen not as the incomprehensible totality of some new world system but as "momentum," physical energy and productive potential.

McElroy's "lucid attention" ("*Apollo 17*" 27) to technological structures may seem superficially like an abdication of imagination altogether, the verbal equivalent of precisionism in painting, with the artist ceding far too much authority to the engineer and to the scientist in creating aesthetic as well as technical forms. In practice the kind of attention McElroy is urging, on himself and on other writers, is a way of ensuring, in a time when nothing is known with certainty, that "true imagination," at least, should "grow from truth." McElroy writes, "The VAB in its own crystal clarity is the inner and visible structure of an outward and limitless conception" ("Neighborhoods" 211), a conception that, precisely because it exceeds any single totalizing perspective, had best be formulated in language that holds close to the observable, with a style that resists the false assurance of an unalienated organic consciousness heroically opposed to some oppressive technology.

By eschewing the preformulated metaphysics and oppositional histrionics that characterize a more aggressive stylist such as Mailer, McElroy stands a chance of learning a style from the engineer. Yet his reason for valuing "clarity" can be quite different from the engineer's, since for the engineer clarity might possess merely a functional value, as the means to a greater economy of force and energy. For McElroy, as for Wittgenstein in his notebook entries on science and culture, clarity is valued not for what it explains or might reveal about a given logical, linguistic, or technological structure, but for the possible conception that might grow from it. In a progressive civilization occupied "with building an ever more complicated structure," Wittgenstein claims, "even clarity is sought only as a means to this end, not as an end in itself. For me on the contrary clarity, perspicuity are valuable in themselves. I am not interested in constructing a building, so much as having a perspicuous view of the foundations of possible buildings" (7).

Although McElroy is quite without Wittgenstein's occasional Old World nostalgia for a premodernist culture, his own creative practice parallels Wittgenstein's concern with knowing the foundations of our thought and with opening conceptual spaces and possibilities of growth out of real structures that are already in place in the world. The writer, like the philosopher, does not take an interest in science merely to know new facts; nor does either one attempt to dismantle or look beyond the structure that science and the languages of science have made available. It is not the writer's concern either to progress with society or to transcend its structure. Instead, one makes abstract connections among facts that are already known, and from this wider perspective readers may look back on the structure and see it anew. (In *Plus,* for example, the mind is made to grow by seeing *through* surrounding "particles to forms that were not beyond the particles" [194].)

McElroy concludes the article on *Apollo 17* by observing that the space program may itself already have translated us "toward forms more cerebral," which is "not necessarily a good thing, by our old standards. But the next thing" (29). In *Plus* he would seek to align his own imaginative forms with the still unapprehended forms within biology and advanced technology. But the "cerebral" form of that novel owes as much to an ongoing struggle with his own work as it does to any particular technological influence. Toward the end of *Lookout Cartridge,* for example, the narrator, Cartwright, connects his own style of storytelling with the *Plus*-like concept of a "Body Brain," which is not "confined to serial single-file one quest-at-a-time circuit-seeking" like the digital computer or narrative drama, but which "sends countless of these single files not one at a time but all at once circulating down the deltas, through the gorges and moving targets and . . . athwart the axes of all pulsing fields" (447). The analog computer, whose parallel processing is now thought to resemble the brain's neurological working better than any linear or causal technology, was a formal model for McElroy's fiction long before he encountered such technology at Cape Kennedy.

Here it is worth recalling, however, Charles Newman's objection to the demands that such computer-generated complexity makes on readers, especially in a novel such as *Lookout Cartridge* which requires constant immersion in a "great multiple field of impinging informations" (*Lookout* 465). We are asked to take it on faith that the characters and systems at work in *Lookout Cartridge* will gather toward some wider scheme; with one of the original reviewers we may experience the narrative as a collection of "subsystems waiting for a supersystem to subsume them" (Stade 279). Yet the difficulty with such an *assumed* coherence is that no one bit of information ever seems to be excluded or excludable. So decentered and relational is each person's experience that the information we need in order to know where we fit in the system at any moment might come from anywhere:

> You have this in your head. You have it from some grainy wire-service photo on the way to the editorial page or the fishing column or the real estate; or you have it moving live contained by your living-room television; or you've had it shown you in the enlarged privacy of a dark theater; or you have it from less pure sources, someone has told you it.
>
> Or, as with me, the image emerged out of standard elements while you sat in a dark projection room during an intermission half-listening to a couple of computer-filmmakers argue whether one can compose more freely with plasma-crystal panels.
>
> You may never be called to account for what you've seen, but you've

seen the glove port and the white coveralls and some of the semi-automated gear on the other side; and you've seen, let us say, radioactive material the glove hand handles without contamination.

Like me, you have in your head things you may not have exactly seen.

Like a lookout cartridge. (*Lookout* 6–7)

It is easy to feel the seductive freedom of such a style, though somewhat harder to know what the style is freeing us to do. The image of the gloved hand handling radioactive material will reappear throughout the book to become, in the words of one critic, "a metaphor for how we can extend ourselves into a potentially dangerous space" (Johnston 96). The camera, the eponymous "lookout cartridge" (a section of film capable of being inserted between other sections), and later the computer will all come to be viewed similarly as technological extensions of ourselves. Yet these technologies can also distance us from experience, to the point where, in the novel, every detail that falls within Cartwright's consciousness has an equal, and equally affectless, significance. The implicit promise, again, is that the details Cartwright observes will contribute to a pattern that must remain immanent, so that his unknowing itself becomes a subject of the narrative. Yet through all his perhaps too perspicuous investigation and ever deepening insight, the best he can do is to *suggest* a world system that no single mind or imagination could ever comprehend.[3]

Finally, the narrative of *Lookout Cartridge* is for me as the music of Charles Ives is for Cartwright himself: "too intelligent, as if some old American strains were interrupting each other so as to break down into their comparative frequencies, so you got their true neural meanings only to find that after all you didn't really want these explicit" (119–20). The novel's mass may have been necessitated by a wholly admirable ambition to present thought's "pulsing fields" in all their social, historical, and economic particulars. But the "splendid chaos" in which Cartwright immerses himself all but thwarts his ability "to make pragmatic as well as ethical decisions" (Hantke 139). The reduc-

3. In a 1985 preface to the novel, "One Reader to Another," McElroy admits that "Cartwright is not going to be enough. . . . He is a window or an eye toward some brink of consciousness." Cartwright's deficiencies—as a man, if not as a narrative agent—are noted by Steffen Hantke: "the appetite with which Cartwright accumulates information keeps him from determining the right measure of how much he can and must know" (138).

tion of complexity that is necessary for any agency is never made by him.

McElroy's epistemological novel ultimately overwhelms the realistic novel of international action, and for this reason the demands of his narrative seem to me less justified than in the novel that followed, in which all questions of agency and epistemology have a direct bearing on the actual growth of a mind. In each case, as in much of his fiction, McElroy would seem to be attempting a simulation or abstract formulation of dramatic devices that compel us forward in our reading of conventional novels, though his is a drama without actors in which human agents are replaced by technological and bureaucratic agenc*ies*. He is translating recognizable generic materials into more conceptual spaces in the hope of evolving toward a new and newly felt concreteness. I believe it is in the shorter fictions, however, and in the more apparently autobiographical interview/essay "Midcourse Corrections," that we find McElroy working closest to his talents, making the purest use of a restricted set of materials, and sustaining, over the length of entire works, the kind of clarity of attention and ratiocinative language that are, for him, key formal resources of an abstract and analytical "drama."

The growth of the book's subject, Imp Plus, into consciousness, and his concurrent growth into a "Body Brain" toward the end, is itself the sole drama of *Plus*, a development that allows neither the author nor the reader to take anything "human" in the story for granted. Anxiety, desire, dreams, creative thought, the sense of the self and its contingency grow directly from the minimal biological, technological, and linguistic situation in which Imp Plus finds himself. This is a growth of consciousness unique in American fiction, but which, for all its strangeness, produces ways of knowing that will be familiar to contemporary readers.

Language and the Origin of Consciousness

In *The Origin of Consciousness in the Breakdown of the Bicameral Mind*, a book that appeared around the time of *Plus*, Julian Jaynes argues that consciousness itself is created through a process of linguistic building: "Language allows the metaphors of things to increase perception and attention, and so to give new names to things of new importance" (138). It is metaphor, Jaynes supposes, that generates new language and creates "previously unspeakable distinctions" to describe new experience (49). As the world grows more complex, we use metaphor—a device that relates things known to things not yet known—to make the unfamiliar world appear more familiar. For example, the English verb *to be*, originating in the Sanskrit *bhu*, "to grow," is itself a metaphor that may have enabled early humans, not yet conscious of their own separate existence, to imagine existence itself as a kind of growth. Similarly, Jaynes notes, "the English forms 'am' and 'is' have evolved from the same root as the Sanskrit *asmi*, 'to breathe,'" though he adds, "of course, we are not conscious that the concept of being is thus generated from a metaphor about growing and breathing" (51). As consciousness advances into ever new areas, its generative metaphors are forgotten and absorbed into the abstract language that we use habitually and automatically.

McElroy does not think of *Plus* as a metaphorical novel, and he would hardly share Jaynes's confidence in the pursuit of historical "origins." But his integral use of language as a constructive technology is consistent with Jaynes's thesis. As in Jaynes, language in *Plus* is made to function abstractly, as "an organ of perception, not simply a means of communication" (*Origin* 50). At the same time, the novel probes the underlying concreteness within the abstractions of ordinary language, thus creating a sense of immediacy and discovery that makes us aware of things we might not otherwise be conscious of knowing. It is one of Jaynes's more dramatic arguments (which anticipates the arguments of Jacques Derrida) that most of what we do, even when we are engaged in "learning" and "thinking," is not done consciously but is given to us by language and by our nervous systems. Or, to cite a French precursor of Jaynes and Derrida, here is Blaise Pascal on the automatism of language—the network of signifiers in which we are caught—in determining the course of our "inner" reasoning: "For we must make no mistake about ourselves: we are as much automaton as mind. . . . Proofs only convince the mind; habit provides the strongest proofs and those that are most believed. It inclines the automaton, which leads the mind unconsciously along with it" (274).

The great originality—and difficulty—of McElroy's art lies in his attention to those habitual aspects of our mental life and everyday behavior that are ordinarily lost to consciousness (so deeply are they rooted in the pre-linguistic experience of our *bodies*). By taking language and perception down to their elements, he makes us conscious of "more," of the unconscious and "unfamiliar" that is always a part of our habitual linguistic reality. (McElroy, incidentally, may be emphasizing how habit "inclines the automaton" by frequently having Imp Plus "*incline* to think" something.)

At the start of the novel, all we have to go on are a few facts and words that, for utilitarian purposes around the spacecraft, Imp Plus was programmed or "prepared to remember" (*Plus* 4) and the equally routine, equally programmatic electronic "impulses" from Ground. Such computer language and "the mission commentary exchanges between SC (the spacecraft) and CAPCOM (the spacecraft communicator in Houston)" had struck McElroy in one review as being "close to . . . an authentic poetry of our period" (Review of Cooper 5). But Imp Plus's thought goes beyond its initiating "impulses from Earth which went on and on," beyond even the formal precision and occasional accidental beauty of such language (*Plus* 10). Indeed, we may begin to sense McElroy's subtler poetic purpose in the play on the sounds "impulse . . . Imp Plus . . . impulse" whose alternation in close proximity throughout the first chapter creates a palpable rhythm that helps suggest the nonverbal character of the brain's early thought. No less than their referential value, the sound and the very *feel* of words are among the forms and rhythms that bear on the brain's growth.[4]

A memory early in the first chapter—of the *Apollo* flights that predated this journey—illustrates the oblique way Imp Plus comes to perceive his identity:

> How long was the start and where was the beginning? The last Apollos had left. But there are no beginnings, Imp Plus thought.
>
> The last Apollos had left the Moon. They had raised their faceplates, lifted off their helmets with the gold sunshades, taken off their hoses and their boots. They had begun not to be new, not needed—though who knew what had got inside them through their suits to stay. Or to go out the other side after blazing a trace inside them.
>
> A trace that would stay in a wall inside—inside what?—in a gradient—but stopped—in the wall of what?—stopped in a cell wall until a pulse could use that trace. But particles that passed thus through bodies might, as through grids, leave nothing.
>
> Could that be?
>
> Impulses on a frequency from Earth raised questions but not that question. Earth had raised Imp Plus and could be rid of him.

4. McElroy speaks of the "visual music in [Imp Plus's] being" in his discussion with the violist John Graham, who had "thought of *Plus* as an opera" ("Midcourse Corrections" 29).

> Something had been taken away.
> From Imp Plus.
> . . . There was a project. (9)

The gradual awakening to self-knowledge in this passage is meant to be a part of the natural and technological processes that go on in and around Imp Plus, a verbal inscription that is consistent with, but not limited to, the imprint left by trace particles in human or material bodies: X rays that "had got inside" the *Apollo* astronauts "through their suits to stay. Or to go out the other side after blazing a trace inside them." While Imp Plus reflects on these material processes, he is able to make tentative mental excursions outward into the "more all around" that he inhabits (12 passim) and inward, gradually, through buried levels of a psyche he can only partly reconstruct. The present memory of the now obsolescent *Apollo* program leads by association to his own beginnings in a similar "project" and, by extension, to the realization that he, too, could one day be discarded. Thus, as consciousness of one's existence implies an awareness of the possibility of *not* existing, the death of Imp Plus (which has in a sense already occurred) is implied at the very "start" of his thought.

Typifying McElroy's style in the novel, this passage features a spare language that seems to grow out of itself in convolutions of syntax, puns, and significant repetitions. The thought of questions "raised" from the control station on earth leads Imp Plus to remember that he too had been "raised" from earth at his launch into space. The *Apollo* rockets similarly had "left," but for the space of a sentence McElroy holds on to the object, "moon," and so allows us to conflate the lunar departure with the rocket's initial liftoff from earth. This conflation of historical beginnings and endings continues with what I take to be the astronauts' raising of faceplates and helmets on *returning* to earth. Without denying the literal, familiar, and referential value of the words "raised," "lift," and "left," and without forgetting the real circuit of historical *Apollo* flights, McElroy in this passage goes beyond objective reference to use the words themselves as a kind of plastic material that advances or shapes the narrative even as it directs the brain's own growth.

Such ambiguity, as well as the low level of "redundancy" in McElroy's prose (which frustrates our attempts to fix the meanings of key words by changing their referents over several repetitions), at once foregrounds the materiality of language and evacuates it of meaning. This disruption of the signifying value of words frees them from the symbolic order that through mechanical or habitual use we tend to construct from them, and teaches us a great deal about the arbitrariness and plasticity of common concepts. The result of such disruption, David Porush notes in *The Soft Machine*, is a dismantling of "grammatical or semantic mechanics" that in *Plus* is "congruent and complementary to the story of the brain rejecting the tyranny of the mechanical role assigned it" (181, 178). There is a linguistic freedom, in other words, that is analogous to the freedom with which the brain directs its growth from strange materials.

Interestingly, Porush associates the style of *Plus* with the technique of "de-automatization," a term from Russian formalist criticism that is often misleadingly translated as "defamiliarization" (176).[5] McElroy's presentation of material in *Plus*, in language that is unexpected and even outlandish, is a way of freeing us from the automatic and control-oriented aspects of linguistic, no less than material, technologies. But I think it is a mistake to regard the style of *Plus*, as Porush does, as primarily nonreferential and "absolutely devoted to making familiar objects and words unfamiliar" (178). Although McElroy can in this way be made amenable to the rhetoric of poststructuralist theory, the denial of reference leaves unexplained much that is humanly felt in the novel. Moreover, the separation of narrative from familiar experience can lead all too quickly to readings that turn the novel into an isolated adventure of the mind wholly separate from sources of power in the world. The true strangeness of *Plus* depends on the continuing pressure of the familiar, the ordinary, the personal life, and the creative struggle that come through in the fiction. Its language invokes a world that we must apprehend as feeling subjects in order for our thoughts and actions to have meaning. To read the novel as a wholly "weightless" and nonreferential linguistic tour de force is to consign its strangeness to

5. The term *defamiliarization*, though widely accepted, is misleading because, as it was originally used by Viktor Shklovsky, the process is "in fact the reverse" of what the term implies. As Benjamin Sher, a recent translator of Shklovsky, has pointed out, the formalist process "is not a transition from the 'familiar' to the 'unknown' (implicitly). On the contrary, it [proceeds from] the cognitively known (the language of science), the rules and formulas that arise from a search for an economy of mental effort, to the familiarly known, that is, to real knowledge that expands and 'complicates' our perceptual process in the rich use of metaphors, similes, and a host of other figures of speech" (Shklovsky xix). What is true of Shklovsky's process (Sher translates the term from the Russian, *ostraniene*, not as "de-automatization" or "defamiliarization" but as the more positive "enstrangement") is true of modern literary art generally. Rather than leading away from ordinary experience and into the unknown, the creative process works toward a fuller expression of what we may not realize we know already.

a critical rhetoric that, while once fashionable, misses the more than literary importance of McElroy's work.

Often, what we know in McElroy's fiction is discovered through a breaking down and recomposition of known elements whose familiar power has not yet been realized. For example, the narrator of *Lookout Cartridge* forgoes direct formulations of even the most easily recognized objects and at times puts obstacles in our way. Like the figure handling radioactive materials through a glove port, Cartwright claims to have "felt things I could see yet with a touch and sight unknown to each other" (134). The narrative, when thus abstracted from the usual sensuous data (such as are taken for granted in naturalistic realism, which would deny its own ideology through one or another "scientific" model), can also lead to a new concreteness in the breakdown, recombination, and endless interpretation of sensual things. In this way technology can itself be a mode of de-automatization, whose power, again, is to deepen and extend our perceptions.

Much of Imp Plus's growth is similarly the result of his "discovering what he knew already," which need not be specifically technological (*Plus* 19). The brain's groping recollection of what he has lost (a literal and at times grotesque re-membering) produces a narrative of fresh and eerily sensual images—"a parchment shine of crisscross called the palm of his hand" (112), "[the] mouth that had laughed a spiral up the grid of his spine, and turned him around" (37), "a salmon-nippled tongue" (75), "the clear-curved skull he did not have" (98), laughter on earth that is "a letting go of something coiled inside something else" (19). But more than the sharpness of the image, what gives affective power to these lost parts of the body is the connection they retain with the mind's present desire. They are all part of what Imp Plus has to get back to (though he cannot), and as he mentally reconstructs what he once was ("mixing memory and desire" more intimately horrible yet never so depressed as in Eliot's *Waste Land*), the memory of his body becomes a focus of desire that both helps and opposes his future growth.

We might reflect for a moment on the use Imp Plus makes of the cluster of words that come to be associated with "laughter," since like other word clusters in the narrative this one suggests the abstract yet sensuous link between verbal growth, desire, and an individual's access to new power. At first the idea of laughter stimulates Imp Plus to think of a woman he was with one spring day by the sea and whom he had met while taking a last vacation from the project. Their car had broken down in Southern California, and Imp Plus now remembers her laughing at the engineer's failure to make sense of the "coils" and "points" of the car's distributor. In the brain's current perceptual field, these fragmentary and unstable linguistic utterances no longer have meaning in themselves. Verbal coincidences surrounding the "coils" (of wire, of laughter) and "points" (on a graph, in an argument, "the point of jokes" [204]) take on as much significance as anything the couple had meant to say or do. Still, the light erotic tone of the vacation seems not to have been lost. The brain has no sensory organs and in its current state does not even know what the word "*felt* was like" (24); "Imp Plus had not known *mouth* or *her*" (38). But even without these words, he can recall enough of the vacation to realize that what he desires to recover is not limited to his own body: "And instead of desired memory to help him into what might lie ahead, he'd been cast back into remembered desire" (37). (Note McElroy's emphasis on all the words that function for Imp Plus as signifiers without reference.) It is perhaps a measure both of the novel's true horror and of how far from cold formalism the narration of the brain's growth can be that "the desire" for another person's body "was a thing not lost" to consciousness (67).

Nowhere is the complex drama of sublime referentiality so powerfully enacted as in this imaginary, wholly linguistic engagement with the Other. In the psychic and linguistic economy that we have come to associate with the sublime, Imp Plus is aware of the reality of a woman's body through her irrecoverable presence, and he retreats from the horror of his situation into a free-floating language. Yet his linguistic reality is not a denial of reference. The body he desires, like any sublime object, is made all the more painfully real to Imp Plus by virtue of its unattainability, and this physically absent, linguistically unrepresentable reality is what sets the mind in motion toward "the more all around," which is experienced as both a filling out of himself and a growing void.

Besides reminding him of the woman on the beach, laughter also makes Imp Plus think of a director of the project who is lecturing in front of a chalkboard. It was this project director, known throughout the novel as "Acrid Voice," who "had said Imp Plus might learn to use the Concentration Loop to talk to himself," as he is now in fact doing (119). Acrid Voice had also once asked, through cigarette smoke and "acrid laughter," whether the engineer would want to "go on forever" in his orbital routine, to which the engineer, in a moment of pique, had responded: "Stick to the point." A second project director, known to Imp Plus only as "Good Voice," had been in favor of an endless continuation of limited monitoring activities (the "impulses" that are now going "on and on" like the plucking of a heavy string [10]). Good Voice wants communication to be restricted to the automatic, positive transfer of information; his concern is to limit Imp Plus's mental activity to a "Dim Echo" of his former thought "that went on communicating knowns by knowns," whereas the consciousness now growing within Imp Plus feels more like "putting unknowns over unknowns" (23). This new consciousness allows him not simply to accumulate knowledge but to name and differentiate aspects of the familiar world, to see "lesser things in the more" (43).

Although the engineer in his illness and helplessness had been inclined to prefer the consolations of Good Voice, "telling . . . of the future and its goodness" (70), to Acrid Voice's perceived "ill will," it turns out to be Acrid Voice who provides him with information that is crucial to the brain's later independence from earth and development in space. Whereas Good Voice, for example, had been wont to mask the probable horror of the brain's existence with banal assurances of the engineer's place in the history of technical progress, Acrid Voice asks him coldly to think of what he is going to lose. But it was also Acrid Voice who had told the engineer about camouflage procedures to hide the IMP spacecraft from foreign predators, but which could also be used by Imp Plus to confuse control from Ground. These and other technical details become of use near the close of the novel, when Imp Plus does in fact take control of the craft's descending orbits to end his flight—and his new life.

Fragmentary political and sensual dramas in *Plus* help to situate the novel in human reality, implying an underlying real-life history that still functions, but in a revised way that is of use in the brain's current nonhuman situation. (Elsewhere, a blind news vendor in New York City, a black nurse with a hypodermic needle, and the engineer with his wife and their child at a beach in Mexico similarly recall Imp Plus's life as a man.) Yet the recollected fragments cannot be said in themselves to contain the meaning of the brain's experience. As Imp Plus learns to think his own growth, the personal and erotic memories recede. He realizes that in space "he could never do what had been done" as a man on earth (44), and if this pained awareness of separation from a past life initially threatens to paralyze him, eventually it stimulates him to discover new channels for his desire, and to accomplish physical and imaginative "migrations of himself" (34) into new situations. The same associations that lead Imp Plus to think of the colors of the sea on that spring day on earth, when he had felt "a desire like the windy flight of long-winged birds shearing the tips of the sea" (28–29), lead him also to remember what he had learned of a biological source of energy, the cell's "mitochondrion, which in great numbers must as a power plant have let off later the laughter which Imp Plus answered on impulse" (30). The word "laughter," no longer a part of the brain's notional network of everyday reality, is thus now of "use" in a conceptual field where human exchanges are equal to the exchanges of biological and electrical power within and through the spacecraft.

I have been using the word *power* in a quite specific sense to refer to the brain's attachment to memories, words, and natural materials whose connective processes parallel the force of thinking. But the brain's growth, specific as it might seem to McElroy's conception, bears more generally on contemporary developments in feminist and post-Freudian thought, whose proponents see the human body as a field in which power relations work themselves out. Like Donna Haraway in her "Cyborg Manifesto," McElroy in *Plus* was learning "how to write the body; how to weave eroticism, cosmology, and politics from imagery of embodiment . . . from imagery of fragmentation and reconstitution of bodies" (Haraway, "Manifesto" 174). Imp Plus is clearly a cyborg, "a hybrid creature, composed of organism and machine" (*Simians* 1). More important, when restructuring his human desires he comes to exemplify the "monstrous world without gender" ("Manifesto" 181) that Haraway hopes for.[6] This world is monstrous in the same sense that any de-automatized world of the imagination is: it runs counter to worn-out (often masculinist) conceptions of time, space, organic growth, and apocalyptic return to an original wholeness. Furthermore, the cyborg does not appropriate the powers of its component parts into a higher unity but allows them to coexist in a nonhierarchical and multiply focused field, whose expression disintegrates differences not only between human being and machine but between mind and nature, male and female, thought and inanimate matter.

What such a world might actually be like for people to live in would become the subject of McElroy's massive novel *Women and Men*, which succeeded *Plus* after an interval of eleven years. The psychological reality of sexual relations comes through no less powerfully in *Plus*, if sex is understood as the contemporary experience that best approximates a sensuous and participating consciousness. Much of the central section of *Plus* narrates the brain's slow reconceptualization of its

6. In referring to Imp Plus, I retain the masculine pronoun as a convenience; McElroy does the same, making it clear all the while that the brain in its later development is genderless.

former desire, to the point where interactions between the brain and the sun have a distinctly sexual character: "Desire had met the Sun. The arcs of lumen and glucose lumen wheeled not from Imp Plus and not from the Sun, but from their mingling that was deeper than touch" (112). Unlike remembered desire on earth, which originates in separation and a lack, this new desire builds into the space that surrounds him—space conceived not as an emptiness to be penetrated but as a "more all around" that desire can join (12). His growth collapses distinctions between subject and object, self and other, language and world, so that for Imp Plus "this describing was being" (204); but this identification of mind and nature is not a romantic narrative of self-realization or transcendence after technological alienation. Imp Plus does not reach back to an innocent oneness with the Mother; nor does he attempt to achieve technological control of a feminized Nature, as does Good Voice, for example, when he speaks of the sun as "a beautiful living bomb of a cow" to "catch and milk" (107). The sexualized relation between Imp Plus and Sun includes technology as part of an animated material reality.

For Good Voice, chief apologist for a dominant and dominating technological society, nature remains something outside and separate, to be manipulated (as the engineer that Imp Plus had been) for utilitarian purposes. Later, when Good Voice discovers an unexpected, potentially world-destroying power in the IMP, his best argument for the capsule's recovery is a catalogue of further functional uses to which such power can be put:

> . . . solar cookers in pueblos, the race for the Reversible Reaction, trans-world power ponds pooling greater areas narrowing greater space, the fight for power which was the fight to find, beyond perfectly imperfect semi-conductors or beyond an element out of which to make wasteless black bodies for solar receivers, the clue to Reactive Reversibility by which to bend spent Energy through the interface of its own use and thence refract it rewound rebound. . . . (189)

McElroy is not altogether dismissive of this utopian desire for an unlimited accumulation of power: he recognizes that such fantasies are often needed to push scientists toward the hard experimental work they do in nuclear fusion, solar energy, and semiconductor and related technologies that promise at least locally to counteract entropy, ecological degradation, and economic chaos. The idealized goal of "Reactive Reversibility," for example, while calling to mind efforts (now considered unrealistic) to achieve high-energy fusion, approximates the reciprocal processes through which Imp Plus and his environment exchange energy. Yet such power is, finally, of minor importance when compared to the epistemological power that Imp Plus has achieved by assimilating himself to language, memories, and physical phenomena, a recovery of external reference after psychic dismemberment and disintegration. At the end of the book it is Acrid Voice, not Good Voice, who receives intimations of this power through a telepathic circuit beyond the range of utilitarian communications, a drawing together of two minds described as something "close to fusion" (214).

While this telepathic discourse is going on between Acrid Voice and Imp Plus, a power play has developed on earth between Acrid Voice and CAPCOM over the recovery of the craft. But Imp Plus will not be "a bone of contention" between them (213). Coming down toward earth's atmosphere, he considers cutting off communications with earth and caroming into deep outer space, but he does not in the end deny the possibility of communication altogether. So that his power can be felt as a presence in the world, he sacrifices his physical existence to become a "circuit of conception," a felt "absence" that will engender further independent growth in the minds of others (215): "The IMP had burned up in the first friction of the atmosphere: thought wondering then what chances now turned upon this fresh absence that would be as lasting as the glint of its arrival must have been brief for any who saw it in the sky: thought wondering, too, if at the last the great lattice had let this happen or had been surprised" (215).

Perhaps the first thing to note about this strange and powerful closing moment is its evident but elusive teleology: Imp Plus moves toward a future that seems somehow determined, but not in any conventional, linear determination that proceeds from past cause to later effect. The passage's odd tense structure enacts a careful reversal of causality by modulating from Imp Plus's ongoing present ("thought wondering") to a past and past perfect in which his *future* appearance to others "must have been brief" and he either "had let this happen or had been surprised." Immanent in all his thought has been a future perspective that seems to have itself determined his past direction (without allowing him to settle finally on one or another possibility), and which includes the possible perception of his own actions in the mind of others.

Such a paradoxical, backward-looking teleology is especially demanding, since it runs counter not only to the linear, causal models of a positivist science but also to the romantic, "vitalistic" teleologies that are typically invoked against scientific models of causation. Yet the paradoxical figuration of Imp Plus's end state is not peculiar to McElroy, or attributable merely to the postmodern novel's preference for endings that cleverly resist closure. Rather, this teleology is wholly consistent with more recent biologically based models in which a system's evolution is indeed in a sense "determined" by a future state. From Aristotle on we have tended to think of teleology in spiritual terms, as a forward-looking "purpose" that seems mysteriously to direct the growth of a system or organism. The physical sciences have uniformly rejected this kind of teleology as a form of religious vitalism, or at best an anthropomorphic projection of human "foresight" onto nonhuman nature. But it may be possible that a different teleology inheres in complex systems, whether conscious or not, living or inanimate, individual or collective (as in the evolution of an organism, and of symbolic and social systems such as a human language or concept). In Ludwig von Bertalanffy's systems theory, for example, "finality" in a biological system need not grow out of a past condition: "Finality can be spoken of also in the sense of *dependence on the future* . . . happenings can be considered and described as being determined not by actual conditions, but also by the final state to be reached. The final state to be reached in future is not a *vis a fronte* mysteriously attracting the system, but only another expression for causal *vires a tergo*" (76–77).[7]

According to von Bertalanffy, only a "closed system" (i.e., one that exchanges no energy or information with its environment) need follow the traditional development from specified initial conditions through a progressive diminishment of possibilities to a single, unambiguous end state. An "open system," by contrast, is not amenable to single determinations, and in this sense Imp Plus must be recognized as the open system par excellence: whether woman or man, organism or linguistic construction, conscious agent or partially conscious cyborg "system," Imp Plus remains open to multiple possibilities and sustains a fluid identity *between* categories. Communication with the outside world continues up to the novel's final sentence, which articulates a condition that is, in von Bertalanffy's systems terms, *multifinal*, capable

7. For this citation from von Bertalanffy, and for many of the ideas in the passages that follow, I am indebted to Christopher Johnson's book *System and Writing in the Philosophy of Jacques Derrida*, especially the discussion of "teleology and code" (145–50).

of leading to diverging, dissimilar end states—to a "surprise," in short, in the imagination of those on earth who may have glimpsed his last traces.

The narrative of *Plus* thus expresses a powerful, nonsatirical, and nonmystical alternative to the great linear narrative of a positivistic, traditionalist science.[8] More generally still, it would be an alternative to any antagonistic drama of tragic autonomy and apocalyptic revelation—although dramatic elements are used to engage our readerly imaginations, and they help us to glimpse last-moment connections that are uncannily similar to traditional devices of closure in the novel. References near the end, for example, to "Christmas Island" (189) in the Pacific Ocean signal a sustained analogy with the birth, death, resurrection, and coming of Christ, but the Christian myth of immortality is systematically negated and recomposed: "He would again not exist but did exist now and would again" (201). As with most inherited forms in McElroy's fiction, the Western salvation narrative has been reformulated to fit cultural purposes that are no longer anthropomorphic. All notions of personal salvation, presence-to-oneself, the Word, and technology as logos do not so much deconstruct as translate into something else, which through some visual music or suspension of disbelief we are to imagine, again, as complexly nonverbal: "The lattice absorbed the words. The words went together into an unsaid emptiness where, having been said more than once before, they were left unsaid. Oh what did that mean?" (195).

This rare emotive outburst is no less moving for expressing a by now familiar problem in the structure of the sublime: the imagination

8. *Plus* represents an alternative to traditional literary histories as well, and it is uncanny in its anticipation of subsequent cyberpunk experiments that imagine a "retro-future" or alternative history determined by present possibilities. Thus Brian McHale, speaking of William Gibson's and Bruce Sterling's novel *The Difference Engine,* uses terms that are wholly consistent with von Bertalanffy's systems teleology: Gibson and Sterling, by presenting a nineteenth century in which Charles Babbage's "difference engine" played the same role as the contemporary computer, have written a novel that "accomplishes for our imaginations the sorts of things that historical fiction once used to accomplish; in particular, it helps us historicize our present by reimagining it as the *past of a determinate future,* just as historical fiction once helped us in a similar way by reimagining the present as the *future of a determinate past*" (McHale 238–39; see also Jameson, "Progress versus Utopia"). McHale would regard the relation of cyberpunk and a work of "high" postmodernism such as *Plus* not as a relation of "influence," and not (as in the case of Pynchon) as a more complex cybernetic relation of "feedback," but as "*independent but parallel development[s]*" (236). Such is the more general relation, I have argued, between the parallel worlds of literature and science, although in a writer as well read in science as McElroy, the parallels are often explicit.

extends its power through the sublime recognition of a system's limitations, in this case the linguistic system that the mind has employed in narrating its own growth. In fact, this sublime recognition would appear to be precisely what has enabled the self to continue building out into the void or plenitude that is not-self, the inexhaustible "more all around" that recurs all through the novel as a nonreferential gesture and so presents to the mind an ever-changing configuration, an inexhaustible significance in the "outside," natural object. Of course, any self-construction built on the ground of its own limitations must ultimately collapse back on itself, as Kant recognized in "Analytic of the Sublime": the mind that "feels itself *set in motion* in the representation of the sublime in nature" is bound either to be "overwhelmed" by an excess of signification or set adrift in an "abyss in which it fears to lose itself" (107). But there is a rare compensation in all this: the recognition of representational inadequacy, though painful, can itself become evidence—or at least a symbolic *figure*—for the entire nonverbal totality, the infinite realm that the finite imagination seeks to apprehend. At this sublime moment of crisis, when the overreaching imagination positions itself at its own linguistic horizon, we have arrived at a new level of narrative signification.

The gambit in which the failure of language becomes a source of renewed signification is of course a powerful rhetorical strategy, one that we have already observed in Pynchon and which we should perhaps not get too readily caught up in. Rather, having come to a "new level of signification" outside the powerful orbit of Imp Plus's thought, we might begin to get something like a critical distance on the extraordinary narrative we have been following. For, in recognizing the imagination's sublime dependence on its own limitations, we come to a point where it may be valuable to ask some hard, skeptical questions of a postromantic ambition to *join with* another which must ultimately frustrate as an impossible desire. To put these questions in terms somewhat unfavorable to McElroy, does not the evocation of an "unsaid," nonverbal reality only give greater poignancy to the self that is able to confront heroically its own linguistic limits? And, to push on to an even more basic question, what agency is it that is capable of perceiving the limits? Should this agency be located within or outside the signifying network, in a relation of immanence or transcendence? And where, now that the question of agency has arisen, is McElroy in all of this, the author and originating voice of the whole?

These questions are taken up in detail in the next chapter, where I consider certain critical and apparently direct autobiographical statements of McElroy's, not in order to isolate some prior or determining Intention of the Author but to better delineate McElroy's elusive construction of a more integral persona, an inscribed author or *compositional self* that enters the text in much the same way that the biographical author would enter the outside, nonlinguistic world. This is not a self that the person "Joseph McElroy" *puts into* his novels, and which the reader can reconstruct from what he or she might happen to know of this person's biography. Rather, like the multifinal end state of a complex system, the compositional self is an identity that discovers itself *in* the text—as an afterthought, so to speak, or the thought that comes to a woman in *Women and Men* "as if she came to it sometime in the future" (479 passim.). It is the self that Don DeLillo has in mind, perhaps, when he speaks of a writer's being able to "know" and "shape himself" through the rhythms and sentence constructions of "his language"; the writer not only sees himself reflected in these constructions "but begins to make himself or remake himself" (LeClair and McCaffery, *Interviews* 82). On the one hand, this compositional identity would appear to be wholly discursive and constituted in language. But on the other hand, since language is a communal property that no writer can wholly possess, it is also an identity that exists outside the writer's own mind, suggesting an independent, nonobjective field of reference. "Of course," DeLillo adds, "this is mysterious and subjective territory" (82). The construction of a self-image that is at once outside and inside consciousness—and finally incapable of representing itself *to* itself—is again characteristic of the paradoxical structure of the sublime.

Before we leave the text of *Plus,* however, we might attempt at least to glimpse the outlines of the compositional self in the "great lattice" or signifying network that comes to be equated with Imp Plus's distinctive consciousness. Through the productive annihilation of "words . . . divided into meaning" (190), the brain has come to inhabit a field of linguistic difference in which contradictory statements and ambiguous words cancel one another's meaning in order to produce new meanings. In this respect the lattice might be said to resemble the structuralist field of language, in which a word has meaning not by virtue of its direct and positive correspondence with an object in the world, but in a negative relation to all of the other words in a language. The tendency in structuralist thought, however (at least in its usual poststructuralist literary transformations), has been to extend this theoretical situation to a denial of *any* positive reference outside the signifying network. It is this nonreferential structuralist world that *Plus* has been thought to allegorize. But the narrative identity of *Plus* is constituted, I have argued, not in the solipsistic continuity of a wholly linguistic discourse, but in the discontinuity between one order of discourse and another, in "the problematic interface with the world outside consciousness," in McHale's words (196).

The lattice, existing independently of Imp Plus's physical being, is independent as well of the materiality of language, although it is of course available to us only in textual form. Hence the need for making contact with some undefined Other, for without this contact there is nothing to prevent identity from becoming wholly self-referential and narcissistic. The self seeks an independent reality outside the linguistic field and (ideally) outside even the author's own consciousness. Reality is sought *in relation;* and meaning, as in the paradoxical closing sentence of *Plus*, occurs only in the spaces between consciousness and world, linguistic system and physical reality, and the complementary linguistic systems that constitute both the "author" and the "reader."

Counterintuitive though such a differential, field-based fiction may be, in its deepest processes it is not primarily faithful to literary conceptions of wholeness or organicism, whether structuralist, romanticist, or otherwise. Field in McElroy is more nearly akin, I have argued, to the systems theory of von Bertalanffy, or to the "field concept" in developmental science that Paul Weiss defined early on as "an abstraction trying to give expression to a group of phenomena observed in living systems. Being an abstraction, we cannot expect [the field] to return more than what we have put into it" (292).

McElroy evidently made use of the field concept in narrating Imp Plus's neurological processes, and those familiar with the biological sciences will be able to trace the developmental field of the brain's purely physiological growth down to the smallest fold or fissure. But more important than the truth, falsity, or currency of any given developmental detail (the concepts McElroy imports from Weiss are at least half a century old) is the model Weiss provided of an abstract and speculative style in the sciences not unlike McElroy's literary style. Looking back on the novel's beginnings in "Fiction a Field of Growth: Science at Heart, Action at a Distance," McElroy credits Weiss with having confirmed "some old images of mine to do with dispersed field-growth." He goes on: "For cells owe their election, Weiss concludes, not to their lineage, to their being derived from a particular parcel of an undivided egg, but rather to the operation of forces to which they become subject during their development. This organization is like a field that exists in the materials through which it acts" (31).

In Weiss, McElroy found both a license and an articulated direction for imagining a growth untied to any single originative cell or center, a growth that might embody a writer's own struggle with human and scientific materials that originated and continue to exist outside any literary frame. For what is a writer's presence in the narrative if not "a field that exists in the materials through which it acts"? Indeed, *Plus* may be the most autobiographical of McElroy's novels in its direct presentation of the compositional self, the writer in the writing whose syncretic activity would enable us to build ourselves back into systems, structures, and rhetorics not of our own creation. A growth as intensely personal as it is scientific or natural, this compositional activity may be the nearest a contemporary writer can come to realizing a collaborative and participating consciousness. Certainly it is in the act of composition, the making and remaking of the self as it matches with the language, that the writer has his or her deepest affinities with technological activity. Whether and how this activity can be made to signify "more" than the particular, limited, and potentially solipsistic consciousness of the individual author is another matter, which I take up in the next chapter.

6

Fiction at a Distance: The Compositional Self in "Midcourse Corrections" and *Women and Men*

I have identified Imp Plus as perhaps the most complete embodiment in American literature of Donna Haraway's cyborg, and so *Plus* might be recommended as anticipating and exceeding in stylistic interest those popular science fiction novels that Haraway cites as evidence of a drift in contemporary sensibility away from older narratives of wholeness, organicism, and romantic self-creation. I have spoken of differential and relational models in the growth of Imp Plus, and of McElroy's ambition to supplant Western narratives of identity and teleology with an authentic "discourse of the Other." But I have yet to conceive McElroy's narrative as a relationship not of pure cognition between the mind and the world outside consciousness, but of active understanding among human subjects. As Haraway herself reminds us, the cyborg, for all its liberating potential, "is also the awful apocalyptic *telos* of the 'West's' escalating dominations of abstract individuation, an ultimate self untied at last from all dependency, a man in space" ("Manifesto" 150–51). *Plus,* as we have seen, is a powerfully compact story of the self reconstituting itself, a being that is neither human nor machine, man nor woman, but most definitely *in space,* where all cultural conflict can be transmuted to a function of language, and external reality (the "more all around" that includes but ultimately exceeds its own representations) is transmuted into the substance of mind. Ego problems for Imp Plus tend to be cognitive rather than cultural, and this has made McElroy's novel appear to be an attempt to achieve imaginative autonomy which is ultimately no less aggressive (despite its Wordsworthian pace and reflective quietism) than Mailer's dream narratives.

If *Plus,* then, anticipates Haraway to an extent, it can also be read as a preeminent postmodern version of the "egotistical sublime" that Keats perceived in Wordsworth's early poetry, a narrative of self-construction which is a "thing per se and stands alone," absorbing all otherness, all possibility of negation, into itself (cited in Weiskel 49, 136). This positive, egotistical sublime, according to Weiskel, "suggests immanence, circularity, and a somewhat regressive resistance to alienation of all kinds" (137). Even the self's desire for relationship is problematic: psychologically, the impossible fulfillment of such desire (one cannot *be* the other) tends to lead to that alternation between identity and absence that Weiskel finds in Wordsworth's autobiographical narratives and that we have seen here, for example, in Franz Pökler's alternation between personal identity and impersonal salvation, in Tyrone Slothrop's cycling between infinite paranoid connections and an anti-paranoid world clearing, and in Steven Rojack's cybernetic alternation between extremes of mundane battle and apocalyptic reverie. McElroy's version of the sublime is not so traumatic as these comparable dynamic versions in Pynchon and Mailer, but it is no less intense and is probably more radical in its nondialectical, nonmetaphorical resolution of the inevitable breakdown in systems of discourse and identity.

Plus is indeed often thought to be a linguistically self-contained and nonreferential fiction, as is shown by the (often sympathetic) comments of McElroy's programmatically antirealist, postmodernist readers. The unusual subject matter of this novel does not especially encourage readings in which the mind's development is placed in the larger context of human society. My own reading of *Plus* has concluded ambivalently by recognizing the felt necessity, but questioning the achievement, of the novel's engagement with otherness. I have now reached a point where it may be possible to escape the orbit of Imp Plus's all-compelling consciousness and so maintain at least a *critical* separation (if not an objective distance) from the linguistic world that McElroy's consciousness inhabits in the later, more apparently autobiographical novels and essays. Like Nathaniel Tarn's Auto-anthropology, this writing aspires to be an extension of the self into the Other, a "thematic study of the self as if it were another, indeed a tribe of others when *je est un autre* is the rule" (*Views from the Weaving Mountain*).

All of McElroy's books, from his very first novel to the critical essays written in midcareer, might be read together as a single ongoing autobiography, a life-shape that resembles less a traditional romantic *Bildung* than a rhizomelike growth of language. "Project yourself into the lives of others," says the organizing mechanical intelligence that inhabits David Brooke and attempts to redirect the composition of McElroy's first novel, *A Smuggler's Bible* (1966): "Project, reorient-ate" (6). A quarter of a century later the very title of "Midcourse Corrections" (1990) would present the form of a literary career as an ongoing collaboration with other people, uncertain forces, and "turns of phrase" that are constantly reaching the self and reflecting it from outside (35). The author thus becomes "a passageway and reflection of strange forms and events even to himself," as McElroy, quoting Nietzsche, says of the main male character in *Women and Men*.

Written for a special issue of the *Review of Contemporary Fiction* devoted to his work, "Corrections" is itself an experiment in literary form that in many ways epitomizes the body of McElroy's writing. "With its 3 inserted interviews, its odd proportions, and its highly colored perspectives of me," McElroy writes in a letter, "['Corrections'] is a hybrid fiction, I suppose. A daydream posing as a document."[1]

In it McElroy does the important work—as any truly innovative novelist must—of creating the critical sensibility by which he will be read, and he begins, understandably, with a warning against the possible misuse of biography in fiction and in criticism. Quite early in "Corrections" he speaks of a reading he gave at a community college near his boyhood home in Brooklyn, where he mildly objected to reviewers calling his most recent novel, *The Letter Left to Me*, "autobiographical. Not unflatteringly—yet as if to identify the author's psychic work and explain its aesthetic result" (12). It is typical of McElroy's thought that his concern here is not to complain of mistreatment by reviewers, but rather to trace the etiology of such neglect and misunderstanding in contemporary habits of reading and in the very institution of professional criticism:

1. The letter is dated August 3, 1989. This is how McElroy described the essay immediately after it was written, before he had completed the short story "Canoe Repair," which goes with the essay and further complicates its unusual blend of genres. Later, in a letter of July 2, 1991, McElroy admits that "'hybrid' is by now the wrong word, much overused word, 'mixed' is better, but something else is needed." Rhyzomatic (after Deleuze and Guattari)?

> For a reader ignorant of the author's life, what is it that sounds autobiographical in the prose itself? (I'm close, I'm asking the right question, I panic and go on, making myself up quite possibly.) Instead, these professional readers look past the words which are audible actions of *this* life on this page, to point to facts of another "text"—a life text—supposed to prove that this work didn't come from nowhere.
>
> Winging it, I swam at thoughts I didn't speak: *were* those "words" really "concrete facts"? and in my novel *Plus* was Imp Plus's first name just an acronym, as the story says, from "Interplanetary Monitoring Platform" (NASA's term) or was it spun off the author's habit of writing "IMP" (= "important") in the margins of books? (12)

By thus resisting a direct identification of the work with his "life," McElroy is not upholding a New Critical belief in textual autonomy; indeed, he is current in regarding the author's life as itself a kind of "text" deeply implicated in the texts of other lives. Neither does McElroy think that the work can be reduced to its materiality (the "concrete facts" of its verbal structure). As literary theory McElroy's argument is strong and elegantly stated. Mikhail Bakhtin said much the same thing when he noted the professional critic's tendency to fall back on "two types of empirical objects": "The text can be reduced to its materiality (a form of objective empiricism), or it can be dissolved into the psychic states (those that precede and that follow it) felt by those who produce or perceive such a text (subjective empiricism)" (Todorov 19, summarizing Bakhtin). Unlike those "professional readers" in a scientific age who are prone to reduce literary texts to material objects, the record of an author's or stimulus of a reader's psychological experience, McElroy would encourage an other-regarding and—in Bakhtin's term—properly "dialogical" understanding of his fiction as a collage of "ghost stories . . . integral fragments . . . breathing relations . . . modifications of language editing the rhetoric of what's inside and not disclaiming faith that the words really rendered things and motions outside—and outside, somehow, *consciousness*" ("Corrections" 13).

There is more going on in the Brooklyn school passage, however, than a precise intuition of contemporary theory. Apart from what McElroy *told* this group of students, the manner with which he directs us (the readers of the essay) to a proper center of interest (the writer in the writing) is not neutral. The description of his performance is itself an intimacy, and it helps to create the illusion that the author is giving us now—parenthetically on the page—the private thoughts and feelings he had held back from his audience while speaking: "(I'm close, I'm asking the right question, I panic and go on, making myself up quite possibly)." Such thoughts are precisely the experience that readers *cannot* share, even in the physical presence of the author at a reading. The supposition that we can share them now, and McElroy's claim later in the essay that the narrative of *Women and Men* was meant to provide access to his "unmediated consciousness" ("Midcourse Corrections" 35), implies a conception of authorial identity that is quite the reverse of popular assumptions about the causal and temporal relation between author and text.[2] McElroy suggests that the mental text, even the presumed "life text," does not precede the work at all but exists instead *in* the work, where the reader might imaginatively participate in the compositional or self-creative effort that went into the life/work's composition. As McElroy repeats with numerous variations throughout the essay: "My work makes *me* up" (18).

He suspects that people go to public readings and read interviews with artists out of a "central fantasy that the artist's work is not the real person," and he resists this popular fantasy "because then it's a short step to seeing the work as being a cover-up" (37). While conducting an interview of his own with Robert Walsh in "Midcourse Corrections," he reiterates his reservations about the entire practice of interviewing artists. He is afraid that the interview may be an attempt to "strip away something" (38), as if the audience wants "to catch this person out as a person when they can't catch the person out in the person's work" (36).

2. Apparently I need to defend my understanding of authorial identity not only against popular assumptions but against formidable poststructuralist concepts as well. The German critic Paul Ingendaay considers my conception of a "compositional self," which I worked out in an essay on William Gaddis, as beneath the level of current debates regarding the concept of an author ("Begriff des 'Autors'" [xi]). By compositional self, however, Ingendaay thinks I mean a collection of personae or "partial-I's" ("Partial-Ichs" [xi]) that somehow mask the biographical author, when I am in fact describing a narrative presence or immanence in the writing. What I actually said in that early essay was that Gaddis existed (in *JR*, at least) as "a compositional medium through which ideas, voices, and cultural debris are transformed (like airwaves passing through an antenna) into significant messages" ("The Compositional Self" 666). I also said (in a line Ingendaay cites [28]) that Gaddis goes about "constructing a coherent personal identity in terms that are quite specific to the aesthetic or technical problem that engages him just then" (657). Both formulations give more agency to the author than would, say, Foucault or Barthes. But my conception of an author's narrative presence is hardly biographical (see my brief discussion of *JR*, *Gravity's Rainbow*, and *Libra* in the section on postmodern naturalism in chapter 7).

Inevitably the example of Pynchon comes up, and McElroy judiciously lets pass Walsh's suggestion that a writer's refusal to be interviewed must come from "fear" or "a desire to be above criticism" (38). McElroy respects Pynchon's reticence, just as he respects the obsessive privacy of Charles Ives, but he "can't believe" Walsh's suggestion that an artist would not want his work to be discussed (38).

Unlike Pynchon or Ives, however, and unlike DeLillo or Gaddis or even so selectively public a personality as Mailer, McElroy would seem to have sensed an opening in the audience's desire simply to know things about an artist.[3] He says he has come to think of the interview as "a genre now, like the preface or the one-act play" (36), and it is indicative of more than his difference from Pynchon that, instead of ignoring or willfully resisting what is after all a fact of contemporary culture, he attempts to reformulate the genre, to fight the "[cultural] flow without herniating it," and to adapt the interview's underutilized power to new and original purposes (17).

No fewer than three long interviews appear in "Corrections," each of which is notable first for the way it inverts the usual relation of interviewer to subject. It is McElroy himself who goes out and interviews people he has known in his life: Walsh, a valued editor of *Women and Men* who had gone on, by a happy coincidence with McElroy's theme, to work for Andy Warhol's *Interview* magazine; Edna Conrad, McElroy's sixth-grade teacher, now ninety-five, "who had taught [McElroy] not only grammar but maps" (9); and John Graham, a well-known violist about to relocate from New York City to Rochester, who had once wanted to make *Plus* into an opera. "Corrections" thus documents an author's ongoing encounters with other people—his readers and in some sense his active collaborators during the creation of the work at hand.

The series of interviews culminates modestly, but with an uncanny rightness, in a meeting with Conrad, who had forgotten McElroy, although he had kept her in mind, more or less, for four decades: "You've participated in my life . . . without knowing it," he tells her. "Except in a corner of your spirit you knew that something must be happening with a lot of people but the teacher can't always know the impact she

3. McElroy is also a bit less scrupulous than William Gass, who complains that "people who never read a poet's work will go to hear him read. They want that presence. They want the gossip. They want what they regard as truly real: the poet's goddamn body. When only his words are real" (LeClair and McCaffery, *Interviews* 165).

has and—" ("Corrections" 53). There is nothing more sublime than the thought that the people one has known, if they are still alive, at any moment must all *be* somewhere, doing things and having thoughts in one's absence. McElroy attempts to literalize that thought by including Conrad herself as a character in the present essay. The interview thus brilliantly demonstrates the conceptual realism of McElroy's sublime aesthetic (even if he ends up wondering, at the interview's conclusion, whether Conrad is "less made up than years ago" [53]).

In many ways the interviews—particularly the Conrad interview—are *like* a fiction: they bring the first-person narrator into contact with distinct characters, and so demonstrate how the author's "life has been made up by others," and how other lives are literally "made up" in one's memory. Bakhtin again provides theoretical support for McElroy's procedure: "The affective weight of my life *as a whole* does not exist for me. Only the other is in possession of the values of the being of a given person" (Todorov 98, summarizing Bakhtin). The interviews allow McElroy to recover his own life values from the possession of "the other," and he also sustains the illusion that the reader, too, is participating directly in the writer's creative life, "making [him] up quite possibly" at the moment of reading (12). Also as in a fiction, the author willingly tries to get outside his enclosed consciousness by locating aspects of himself in others, thus ensuring, in Bakhtin's characterization of a polyphonic narrative, that the author's own life story will be made up by "a plurality of consciousnesses, with equal rights and each with its own world," rather than by "a single authorial consciousness" in "a single objective world" (*Problems* 6).

McElroy cites Nietzsche on the body as "a social structure composed of many souls" ("Corrections" 17). Yet he also understands what several postmodern interpreters of both Bakhtin and Nietzsche often fail to realize: that a recognition of plurality does not in and of itself ensure authentic community or self-coherence. McElroy's dialogism, like Bakhtin's, does not take plurality for granted; rather, it would actively *construct* a plural self/community "in a working conversation between 'outside' documentable detail and an inside greedy rhetoric of impulse: still more, an inertial field extending a function of the enclosed perceiving subject outward to describe and establish where I am not (and where the precincts of the controlling author might be given up)" ("Corrections" 17).

Again, and crucially, McElroy locates his compositional self in the space *between* plural subjectivities—a space that he has consistently sought to occupy in all his narratives, from the bridge passages of *A Smuggler's Bible* which collect, correct, and "reorient-ate" the eight "principal parts" of David Brooke, through *Lookout Cartridge*'s mechanical "inserts," and on to the interchapters explicitly titled "BETWEEN US," "BETWEEN HISTORIES," in *Women and Men*. The aesthetic of the novels could not have been more appropriate to the interview "genre" which McElroy recognizes and reinvents in "Corrections." Interview and fictive experiment each affirms the Bakhtinian conception of narrative as capable of encompassing reality, but not because it follows any single consciousness or rigid system. The novel, rather, is "plasticity itself . . . the only possibility open to a genre that structures itself in a zone of direct contact with developing reality" (*Imagination* 39). The only transcendent consciousness of McElroy's later fiction is a consciousness that transcends *into* others. As in the conceptual naturalism of Don DeLillo and William Gaddis, which I consider in the next chapter, the author's consciousness loses itself in the world and comes to form "a community," in the collective words of *Women and Men*, forever in flux and "capable of accommodating even angels real enough to grow by human means" (11).

Taken together, the interviews in "Corrections" embed McElroy's creative life, the life that manifests itself at this moment in this text, in personalities outside the self. But no less important is his grounding of such polyphony not only in other people but also in documents, significant objects, streets he moves through, and rooms he inhabits in the city. Like his stories, "Corrections" becomes a "mixed narrative" (54) designed to "interrupt, interleave, break diverse kinds of document" (10). In this particular narrative, perhaps more than in the novels themselves, he trusts to his own partly directed, partly accidental motion through the city to yield up a life shape that is outside his novelistic control: "It's a writing day, after all, and we're on the run, it's the city uncentered running itself through us as if *we* were not what's in motion" (16–17). In passages bridging the interviews, we are made to follow McElroy's thoughts—or thoughts given to the protagonist who is identified with McElroy's "I"—as this protagonist moves literally by various means of physical transportation: by subway to and from the Brooklyn school reading (where he learned of Conrad's present whereabouts in Vermont), by bicycle through lower Manhattan to meet with Graham, and then on to the midtown offices of Walsh's magazine. He mentions stopping en route at a corner telephone to hear his wife's recorded message, her voice "[coming] into the void but with her imagination's truthful nearness, reporting that she got the test results and she *is* as we thought, pregnant. A plausible fiction. Who am I, at fifty-eight, to be a father again?" (33)

The external, apparently unmediated "life text" again appears to have become a part of the narrative, creating an intimacy more natural and complexly personal than the most explicit confession would allow for. Style is not, as in Adams and other classic modernist writers, a mask or technological displacement of the elusive authorial self. Rather, we are invited to observe the self that is being created at the moment of composition, a moment not to be confused with the self-reflexive, narcissistic engagement with language that is often associated with nonreferential practices of a wholly literary postmodernism.

Of course McElroy, like any other novelist, begins to construct a self-critical authorial presence in "the plurality of languages, discourses, and voices, and the inevitable awareness of language as such" (Todorov 66, summarizing Bakhtin). More fully than most contemporary novelists, McElroy abstracts these various languages into a self-critical linguistic system—basically a *style*—of his own. But it is a system that constantly looks to incorporate into itself other people, other systems and personal styles, somewhat as people in New York are said to have "incarnated and incorporated" Grace Kimball, a major character in *Women and Men*, "into their systems," or as a place—Ship Rock, New Mexico—is said to be "not any place except what's happening around it" (*Women and Men* 129, 91). All positive, self-creative systems and structures in *Women and Men*, whether social, psychic, or natural, are thus defined in relation to what they are not, and in this way they *open out* into the surrounding environment despite their self-reflexive structure. No mere literary theory of "difference" (which might respect and from a distance accommodate any number of unassimilated Others), McElroy's style takes other styles and personalities into itself.[4]

Such contact is continually achieved throughout McElroy's later narratives, whose naturalness and personability is composed as much by the inanimate life of the contextual city as it is by the author's literary activity. "The City, which generated its own noise and change,"

4. McElroy's private but outward-leaning style has perhaps its closest theoretical parallel, again, not in literary theory but in systems theory, where any self-referential structure of a system must remain asymmetrical, open to the environment, and capable of combining with it, so that "the self-referential closure becomes the means by which contact with the environment can be established" (Schwanitz 272).

is actually said, and concretely *shown*, to "think for" one of the minor characters in *Women and Men*, a process no more abstract than finding your way home "by a route so invariable that the apartment house will come to you" (478, 487). Particular media of physical motion or communication—subways, streets, bicycles, telephone answering machines, or electronically coded "moneys finding their way into a faraway bank like a corporate thought"—are inseparable from whatever private event or perception is being narrated (*Women and Men* 201). These media are never neutral, nor are they reducible to the terms of a McLuhanesque "message (read literally *massage*)," which implies a one-way communication between a sender and a passive receiver (*Women and Men* 14). Media in McElroy are instead active conduits that *constitute* the senders, receivers, and all conceivable conversations between them. As the narrator of *Women and Men* says, in a voice that includes the book's readers along with the overheard voices of characters in the novel, "we took place not just in the receivers of our waves of relations but *as* those receivers no less" (13).

This co-implication of medium and message, discourse and referent, is more than metaphorical because language in McElroy is not separate from the world but is a technological extension, articulation, and/or revision of it. Few writers have been capable of moving so generously through the various media of communication without attempting to master them. (Gaddis in *JR*, alone among American writers, and Walter Benjamin in the unfinished *Arcades Project* come to mind as possible comparisons, the modern writers who persist in the attempt to bring technological phenomena to expression.) And few writers could let emerge "between voices" and from out of the "massed phenomenal ambiguous differential exactness the perhaps after all swampy interrelatedness of all things, effects, relations, none distinct by itself" ("Corrections" 17). Only through immersion in material that cannot *be* controlled might the author "know who, as an artist, I am (in so many words)" (17). The form of a literary career, like the backward-looking teleology at the end of *Plus*, is other-directed, multifinal, known retrospectively: "Like the personality of the writer which can be seen in a letter written at the age of eight, you look back on it forty years and you feel you're the same person, maybe it's genes pushing, some desire that never wavers" (30). Narrative in McElroy is the Wordsworthian form of a life actively recollected, a pattern that does not just include or make room for material drawn from the author's private history. The pattern *constitutes* a life that includes the author's search for a pattern: "A serious push, even in the incipient clutter."

I cite so many of these remarks about the linguistic technology of self-creation, the self's efforts to overcome recalcitrant material and take the material into itself, because they suggest an important way in which McElroy's self-consciousness differs from that of his more celebrated American contemporaries. "I am always worried that my fiction would seem to arise from an aesthetic that would seem to be linguistically self-contained," he told Bradford Morrow (159). It is this inward tendency that McElroy at midcareer would appear to be making "corrections" for, as Pynchon in the introduction to *Slow Learner* felt compelled to correct for what he perceived to be an excessive abstraction in his early writing. Like Pynchon, McElroy in more recent work (such as the coming-of-age narrative of *The Letter Left to Me* or "Night Soul," the quiet story of a father listening to his child's troubled breathing) would seem to be emphasizing domestic life as an immediate source of authenticity, "found and taken up," in Pynchon's words, "always at a cost, from deeper, more shared levels of the life we all really live" (*Slow Learner* 21).

Unlike Pynchon, however (at least the Pynchon of the quasi-autobiographical essays written years after *Gravity's Rainbow*), McElroy does not reject "abstraction" as a way into these "deeper," more personal everyday levels. Pynchon, moreover, came to family life relatively late (and in possible ironic reaction to the frantic promotion of "family values" in the last years of the Reagan-Bush era),[5] but the attention to domestic life has been a part of McElroy's work from the start, never separate from his abstracting tendencies. Indeed, McElroy recognizes that such tendencies, like the observantly abstract methods of science and technology, have the power to take us out of the protectively rational self, into the material life of the city, into specific private lives, and into the life and work of an author's peers and precursors. All are made to contribute to the "sum" of his parts and "effects": "I visualize other writers—are they doing that sum?—writers whose names begin with B, with G, with R, with M, with A, with W: I see '(Y/N)Y' at the bottom of my screen and think that I need to be more smartly pessimistic, at

5. Set at the start of Reagan's second term in 1984, *Vineland* is shot through with references to the family, which is invoked by the administration to mask a systematic reduction of civil rights. Thus, when Federal Agent Hector Zuñiga asks Zoyd Wheeler the apparently innocent and meaningless social question, "How about your kid, then?" Zoyd responds with characteristic paranoia: "What about her? I really need to hear some more federal advice right now about how I should be bringin' up my own kid, we know already how much all you Reaganite folks care about the family unit, just from how much you're always in fuckin' around with it" (30–31).

least try" ("Corrections" 17). Although no writer is specifically named here, McElroy has spoken privately of the "pessimism" of Gaddis and Henry Adams, and one could imagine the same being said of Pynchon and Coover.[6] McElroy is at once more direct than these writers in his inclusion of an overspilling "front of private life" and more resourceful in showing how concrete, lived social experience is always caught up in abstract systems, not the least of which are the developing, never completed systems to be found in the writer's own work (*Women and Men* 266).

He knows that the inclusion of direct biographical information is but "one more designing act," but even so, he would let the final design reveal the unfinished quality of his living material (15). He presents life itself in technological terms, as "a lode that will look partly unmined" or as "the provisionally finished work with its sources stirring and even showing" (15). This seems admirable to me, although it might be objected that the realism McElroy is attempting here is fundamentally impossible. By this point in his career, he is seeking both to present his "unmediated consciousness" *and* to project himself into a network of connections outside his own perceptions—that is, to be both inside and outside his self-representation. McElroy, in a repetition of the classic predicament of the sublime imagination from Adams on, has reached the theoretical limits of his ongoing attempt to create an intersubjective biographical identity. For, as Tzvetan Todorov points out in another context, "Even if the author-creator had created the most authentic autobiography or confession, he would nonetheless have remained, insofar as he had produced it, outside of the universe that is represented in it" (52). The mind stands in the way of its own representations; it is itself the obstacle to participation in a communicative network, and to suppose, as McElroy does in "Corrections," that "the obstacle in the way *is* the way" may be wishful thinking (14). Indeed, so much and so articulately does McElroy in his critical writing protest on behalf of a nondialectical communicative network that we may again suspect whether this network, an authentic discourse of the Other, is in fact achieved in the later fiction.

The polyvocal narrative of *Women and Men*—of a scope beyond what any other contemporary writer has attempted—is clearly designed to give up the precincts of the controlling author. (Before *Harlot's Ghost*,

6. "I suspect my sense of failure is as thoroughgoing as Gaddis's or Henry Adams's, but [I am] fighting this with moods other than theirs, and hopes that maybe they gave up on" (letter of January 23, 1991).

Marguerite Young's *Miss MacIntosh, My Darling* was its only rival for sheer length; it outdistances Gaddis's *Recognitions* by a full two hundred pages.) Its meaning aspires to be a tacit construction not even available to the author, and not available to the reader either, but available only (as at the end of *Plus*) in a telepathic relation between the two. It will not be possible—or perhaps necessary at this point—to describe McElroy's novel at length; it is "a simple book and all of one piece," as Harry Mathews notes (199). Unlike "Midcourse Corrections," whose interviews ensure that many distinct "voices" come together to "make up" a multiple authorial consciousness, *Women and Men* employs an inclusive consciousness that moves on occasion into the consciousness of an isolated individual (the protagonists James Mayn or Grace Kimball, for example, or any of a hundred lesser characters), but always merges again with one collective voice: the "we" of the Breathers sections, "whose growing voice breaks into many voices we have always known" (*Women and Men* 19). Like the "breathing relations" that etymologically inspire McElroy's narrative, the forces of language in the novel strive for an organic reciprocity. These forces are at once private and public, "centripetal" and "centrifugal," in Bakhtin's terminology, and kept in tension so as never to ossify into a fixed "univocal" system (*Imagination* 272). But the overall tendency of the narrative, in my experience of it, is decidedly centripetal. Even its sublime, "angelic" register is inflected toward the private authorial center, which continually reduces its "near-angel voices (awfully like ours on a good day) to now and then one voice" (*Women and Men* 27). True to Mathews's only apparently wry observation, but in a manner less favorable than Mathews meant the comment to be, the book does somehow remain single and of a piece.

Anticipating criticisms of the Edna Conrad interview in "Midcourse Corrections," McElroy admits to dominating the conversation: "Say I put my private thoughts in all through our talk, what could I put in for her?" ("Corrections" 53). Conrad is ninety-five, after all, and we can believe McElroy when he claims that the rapport between the two was largely nonverbal (and so lost to the interview transcript). But McElroy's great novel also has the feel of a one-sided conversation. There is much here that is public and wide-ranging, but, despite the ease with which the narrative courses through the various corporate voices in America, and despite the angelic insistence that "there's private life and public life and always was," the narrative keeps coming back to private life (*Women and Men* 253). McElroy writes about real events (such as an *Apollo* launch), but as if from a distance, "acting at long range" (256). The sublime moment comes, most often and most successfully, in private, even intimate, conversation:

> "Do you ever feel," she wonders, "that we fit into a large life that doesn't much know us but—holds us? And that this is better than its being more aware of us?"
>
> "Well, let's not tell it about us," he seems to agree, and she puts an arm on his shoulder and frowns.
>
> "It is beyond understanding us," she pontificates softly. (266)

The unnamed totality is "beyond understanding us," not the other way around, and this couple recognize and preserve their freedom by keeping at a distance from the power that includes them. The male in this conversation says he thinks he has "to go and ask it a few questions" (266), and so articulates the author's desire to extend the private conversation that is *Women and Men* to a larger, more structured system of articulations. The extension to the public sphere is not made, however, any more than Pynchon could have written this moment of sublime intimacy. The desired discourse of the Other remains a more or less private affair, and *Women and Men* does not finally engage the problem (which is by no means McElroy's problem alone, but the general problem of postmodern discourse) of the loss of a public sphere.

I should confess that the critique of *Women and Men* that I am all too sketchily setting out—indeed, the very phrase "discourse of the Other"—is not original with me. For in trying to articulate my reservations about a book and an aesthetic that have both moved me and frustrated comprehension, I have been following an earlier critique of a similar—and similarly equivocal—fiction of identity and absence, inwardness and linguistic growth, vast and revolutionary events imperfectly resolved in the mind's private sphere. Like *Women and Men* (and *Plus* before it), this earlier poetic fiction reads "the past as if it were the future and the future as if it were the past, for memory and desire are linked as derivatives from . . . vacancy" (Weiskel 144). Its governing consciousness would construct itself from a void, but *in relation* to others, so that the fiction has often been read as a prime example not of Bakhtinian dialogism but of "conversation." The conversation is not, however, a mere

> "communication" (the cant word of our social world); its aim is not the transmission of knowledge or a message but the springing loose of an efficacious spirit which haunts the passages of self-knowledge,

> however shallow or deep. Yet to describe [this work] as any kind of conversation seems perverse. Its apparent form is closer to monolithic monologue; it drifts, gets lost, peters out now and then, and generally proceeds without the dramatic constraints a stricter form or a genuine auditor would compel. (Weiskel 169–70)

There could be no better description of *Women and Men* than a "conversation" that needs no second partner, "a discourse apparently exempt from the veridical testing conversation normally entails" (Weiskel 170). But this passage is not, of course, a critique of *Women and Men* but of *The Prelude* by Wordsworth. But this is probably a parallel that I should not pursue at the moment. Instead, I close this chapter by merely suggesting that Wordsworth's romantic model, though the very model of a sublime *failure*, may ultimately prove more appropriate to the critical understanding of *Women and Men* than either the modernist or postmodernist models by which McElroy has most often been measured.

7

From the Sublime to the Beautiful to the Political: Don DeLillo at Midcareer

Don DeLillo has been celebrated as a writer's writer ever since he began publishing fiction in the early seventies, but only recently has he become a possible subject for academic literary criticism. In 1987 Tom LeClair's *In the Loop* succeeded in placing DeLillo among older, more academically established contemporaries such as Robert Coover, William Gaddis, and Thomas Pynchon—the writers who, in DeLillo's own estimation, "set the standard" for contemporary novelists.[1] Yet DeLillo's success has been somewhat different from theirs. At least since the publication of his ninth novel and first bestseller, *Libra* (1988), he has reached a wider and more popular audience, and his chief promoters within academia are those who, like Frank Lentricchia, share his critical engagement with "particular American cultural and political matters" that have made his writing count "beyond the elite circle of connoisseurs of postmodernist criticism and fiction" ("Citizen" 241, 243).[2] Sharing the ambitions of Pynchon, Coover, Gaddis, and even so

1. DeLillo made this remark in a *New York Times* interview with Robert Harris (26; cited in LeClair, *Loop* 22).

2. Thus Lentricchia introduces a special issue of *South Atlantic Quarterly* (Spring 1990) devoted to DeLillo. The contributors, knowing the political requirements of contemporary canon formation, have established DeLillo, first and foremost, as an exemplary *cultural* critic. LeClair, too, had distinguished DeLillo from Pynchon, Gaddis, and Coover as "a writer who works more instinctively, internalizes more fully his cultural context" (*Loop* xii). In England, Malcolm Bradbury, writing in 1992, was purposely looking beyond the American "experimental" novel of the sixties and seventies when he noted that DeLillo is "widely thought to be one of the best and most representative novelists of the 1980s" (18).

reputedly "difficult" an author as Joseph McElroy, DeLillo is the writer who has carried their aesthetic, with the fewest compromises, into the mainstream fiction of the eighties and nineties.[3]

The passage into the present has not been easy or direct, however; nor has it been helped along by DeLillo himself. Far from courting popularity, he has actively sought a position at the margins of cultural power, in self-conscious reaction against those nonliterary media of entertainment and communications that are increasingly setting the terms for the dissemination of knowledge and public discussion of literature. Against "the age and its facile knowledge-market," DeLillo has consistently defended the necessary "difficulty" of modern narrative. In his novel of 1976, *Ratner's Star,* he went so far as to imagine a writer of "crazed prose . . . [who is] working against the age and so he feels some satisfaction at not being widely read. He is diminished by his audience" (LeClair and McCaffery, *Interviews* 87). DeLillo is not himself "this writer," not the high-modernist cultist or the Barthesian author of *illisible* texts; but he does feel the attraction of writing this way, and he risked moving free of public taste in the very book, *Ratner's Star,* that most resembles such seventies "novels of excess" and experimental exuberance as Pynchon's *Gravity's Rainbow,* Gaddis's *JR,* Coover's *Public Burning,* and McElroy's *Lookout Cartridge.*[4]

The ambition of *Ratner's Star* is surely overreaching: not only does the book enrich contemporary fiction with "a cult history" of mathematics, but it also aspires to *be* "a piece of mathematics" (LeClair and McCaffery, *Interviews* 86), to achieve a complex verbal precision and to pattern its two complementary halves on abstract numerical relations (such as the relation of the number 1 and its "imaginary" counterpart, the square root of minus 1). But so long as fiction remains tied to a verbal medium, it cannot hope to present the reality of numbers, except as an absent referent, a metaphor for all the "mystery and wonder"

3. When I call DeLillo a "mainstream" writer, I am thinking of Coover's definition of the "true mainstream." Mainstream fiction, according to Coover, may find its materials in the "priestly" novels of serious realism and in various "folk" genres. But Coover's mainstream does not rest in either of these more conventional forms; it "typically rejects mere modifications in the evolving group mythos, further surface variations on sanctioned themes, and attacks instead the supporting structures themselves, the homologous forms. Whereupon something new enters the world—at least the world of literature, if not always the community beyond" ("On Reading" 38). Brian McHale may have a similar distinction in mind when he calls the conservative wing of mainstream writing "bestseller" fiction and its progressive wing "advanced" or "state-of-the-art" fiction (227).

4. The phrase "novel of excess" is LeClair's in *The Art of Excess: Mastery in Contemporary American Fiction.*

that words *cannot* express (*Interviews* 86). Within the actual narrative a parade of scientists and DeLillo's twelve-year-old protagonist are seldom shown at work, for the same reason that Robert Musil's mathematician protagonist could only rarely be shown doing mathematics in *The Man without Qualities.* As Musil's narrator comments, "nothing is so difficult to represent by literary means as a man thinking" (128). In the absence of this scientific content, *Ratner's Star* itself frequently and deliberately degenerates to static parody which exposes not the mystery of mathematics but the sheer boredom of pedestrian, professionalized, and routinized science.

In fact, DeLillo is not writing about what happens when a scientist thinks; he is writing about the collective activity of two thousand scientists. Bureaucratic power completely overtakes abstract knowledge ("a little boy's pure mathematics," as Piotr Siemion says in "Whale Songs" 52). But a global bureaucracy is scarcely more likely than mathematics to produce narrative. At its best (where it possesses a conceptual subtlety and stylistic energy that anticipates the later, major work), DeLillo's novel becomes one more doomed attempt to present the unpresentable; it becomes, with an interesting mathematical twist, yet another utopian version of the technological sublime.

There has always been something quixotic about all of these infinitely accurate, infinitely complex fictions of the seventies whose ambition cannot rest until it has plumbed every level of an obsession, and whose subjects, taken together, cover the range of institutional languages and contemporary technologies. The authors of these narratives have felt a compulsion to innovate and experiment with fictive form, "to advance the art," as DeLillo said not long after finishing *Ratner's Star*. "Fiction hasn't quite been filled in or done in or worked out. We make our small leaps" (*Interviews* 82). Yet an aesthetic that too readily devotes itself to technical innovation always risks sudden obsolescence, and DeLillo, arriving a mere decade after this first generation of "Mega-Novelists," would soon become more modest in making his "small leaps" and more aware of the limits of a purely aesthetic innovation.[5] The critical injunction to "be experimental" can be as much of a trap as any other creative prescription and can easily lead to that "monotony" which, in the words of a French avant-garde writer of the thirties, "is nourished by the new" (Jean Vaudal, cited in Buck-Morss 104).

The American experimental novel of the seventies could never hope

5. The phrase "Mega-Novelist" is Frederick Karl's in "American Fictions: The Mega-Novel."

to keep pace with technological innovations and consequent changes in the culture at large. Even as *Ratner's Star* was being written, writes Siemion, "the all-enveloping institutional reality DeLillo satirizes, and sustains in doing so, was undergoing a radical shift (and necessitating the creation of some new, more pertinent aesthetics)" ("Whale Songs" 89–90). Siemion refers to the shift away from the single all-consuming discourse of "big" science and "the Keynesian superstate . . . toward deregulation, decentralization, and globalization," and toward a culture of media simulations more thoroughgoing and invisible in its effects than anything that even these writers may have imagined in their most paranoid moments. The new nonhegemonic, multiple, and politically fragmented culture of the eighties would require for its expression something besides the broad-based institutional satire and mathematical abstraction of *Ratner's Star*. Even the "irredeemably heterogeneous texture," the postmodernist "montages of tones, styles, and voices" that Lentricchia finds characteristic of all DeLillo's novels, would not be enough to guarantee either their political legitimacy or their aesthetic relevance ("Citizen" 239–40).

A number of critics besides Siemion have sensed that DeLillo's fiction marks the end of a cycle in contemporary writing. John Kucich, writing of the general "plight of the white male writer" in America today, finds DeLillo's first eight novels hampered by "postmodernism's nonrepresentational aesthetics, which is in some ways part of the legacy left by modernism," and by a politics that locates all conflict "in the structure of language (and in its uses—ideology, cultural codes, modes of intellectual production, etc.), rather than in overt forms of repression, in individuals, or in the shape of historical events" (329–30). (Let it pass that Kucich never ventures to say what it is, if not language, that constitutes individuals and gives shape to history.) The tag ends of literary modernism manifest themselves in "DeLillo's strained form of narrative purity," in an "elaborate narrative detachment" that pretends to be outside power, and in an apparent loss of reference to all but the most topical, already mediated images. The result, according to Kucich, is that neither the author nor his characters can achieve any "active oppositional stances" to the hyperreal mediated culture in which they exist (336–37). Similarly (though from a cold war liberal standpoint opposed to Kucich's rear guard realism), Thomas Schaub has criticized DeLillo (again in the first eight books) for a too easy acceptance of a ready-made culture, for a mesmerization with vast corporate and technological "systems" that causes him to confuse an existential "mystery . . . with systematic manipulation" ("Don DeLillo's

Systems" 131). By seeming to forget that there is a historical basis and political motivation for the current multinational world power—"the GMs, ITTs, IG Farbens, and General Electrics"—DeLillo creates a complicitous fiction that can produce in readers a "contentment before the mystery of existence" and a false sense of "what is now natural" (131).

DeLillo himself would seem to have joined the company of his critics by letting his midcareer fiction be largely *about* the impossibility of achieving a wholly literary opposition. Both of the novels that concern me here, *Libra* (1988) and *Mao II* (1991), center on protagonists who fail in their attempts to merge with history, while DeLillo himself in these books is more than usually concerned with cultural material that escapes literary control. In *Libra* he narrates the life of Lee Harvey Oswald and the assassination of President Kennedy with a documentary realism that forgoes any but the faintest irony and external commentary by the author. In *Mao II* he imagines the emergence of a reclusive writer into the political world, whose crowded and conflicted populations remain immune to whatever language this writer, the standard-bearer of an endangered "species," might bring to them (*Mao II* 26). In the two books, therefore, DeLillo enacts first a figurative and then a literal death of a peculiarly American phenomenon: the great single and disappearing author at the heart of innovative contemporary literature.

In taking things to such extremes, however, DeLillo may also have glimpsed the aesthetic that eluded the crazed and diminished writer of *Ratner's Star*. It is certainly a more modest aesthetic than that of Pynchon in *Gravity's Rainbow* or Gaddis in *JR*, an aesthetic that is not so much oppositional as it is integral to the culture, and not so much sublime as beautiful. Kant distinguished the category once occupied by the beautiful vis-à-vis the sublime in nature thus: "For the beautiful in nature we must seek a ground external to ourselves, but for the sublime one merely in ourselves and the attitude of mind that introduces sublimity into the representation of nature" (93).

With the postmodern death of the author and a consequent loss of a "ground merely in ourselves," DeLillo has sought an external ground, not "in nature" but in the very media and technological systems that have supplanted the natural world as a sublime object of contemplation. To arrive at an aesthetic appropriate to both the postmodern condition and his own talent for self-effacement, he has found it necessary not, primarily, to elaborate further innovations in the manner of McElroy, Coover, Pynchon before *Vineland*, and Gaddis still in the tellingly named *Frolic of His Own* (1994). Nor does DeLillo seek, like Mailer before *The Executioner's Song*, an oppositional ground within himself.

Since *Ratner's Star*, if not before, his approach has been to rework, from the inside, technological forms and political narratives that the culture has already constructed.

To a degree, then, DeLillo's fiction does require us to countenance as "natural" many forms that, in pretechnological times, would never have seemed so. DeLillo has always been a writer for whom the "natural world" is one of created objects, and his fiction has always accepted the collapse of distinctions between nature and technology, and a consequent primacy and proliferation of reproductive images. But we need not accept such conditions passively. Traditionally "the critical, 'negative' power of art" has tended to "[melt] away" in the face of the beautiful (Weiskel 46), and it has certainly been easy for the postmodern imagination to lose itself in the fair world of media appearances. With the loss of the separate and resisting self we have come to associate a loss of critical distance. But DeLillo's achievement may rest precisely on his uncanny ability to contest power from the inside, to unsettle its enabling languages without for a moment lapsing into irony or (as in *Ratner's Star*) a satire that ends by sustaining what it critiques.

The aesthetic of DeLillo's later novels is, in a sense, a step back from the radicalism of *Ratner's Star* and its models in American high postmodernism, but it is no less sophisticated than this earlier work in its use of technology. Instead of reducing multiplicity and media simulations to the equivalent depthless forms of postmodernism, DeLillo takes advantage of technological constructions and uses them to express what he calls "secrets of consciousness, or ways in which consciousness is replicated in the natural world" (DeCurtis, interview 299). The material world becomes, in his fiction, the concrete material and outward form of everyday thought. Technology pervades the most ordinary existence, and by integrating technology into his narrative, DeLillo carries his fiction beyond the limits of a mere literary experimentation to what we might call a postmodern or conceptual naturalism.

Postmodern Naturalism: *Libra*

Early in *Libra* DeLillo presents a memorable caricature of his own situation as an author who must sort through masses of historical data for elements of character, plot, thematic content, and narrative possibilities. The fictional Nicholas Branch, a retired CIA analyst who is writing a "secret history" of the assassination of President Kennedy, faces the same difficulty as the novelist who would represent the large contemporary forces that shape American culture. Branch's materials—the Warren Commission Report and accompanying volumes; films, photographs, medical and legal records, "the data-spew of hundreds of lives" (15)—are certainly materials that DeLillo made use of in writing *Libra*. And the unlikelihood of Branch's ever finishing the project, his despairing suspicion that the history "is the megaton novel James Joyce would have written if he'd moved to Iowa City and lived to be a hundred," suggests the compositional difficulties that DeLillo himself had to overcome in order to write the novel at all (181).

But DeLillo is not Branch, and his work, though culturally ambitious, is not meant to be "the Joycean Book of America . . . the novel in which nothing is left out" (182). For Branch sitting "in the book-filled room, the room of documents, the room of theories and dreams" (14), unrestricted access to so much material promises a total understanding that remains always beyond him, while the still accumulating products of his research remain "lost to syntax and other arrangement" (181). DeLillo himself recognizes that, as long as we remain passively within the historical record, we are going to be "constrained by half-facts or overwhelmed by possibilities." By now the background is too vast and confusing, too wrapped up in private mystery and official secrecy for a novel on the assassination to be anything other than the "refuge" DeLillo intends this novel to be ("Author's Note" to *Libra*).

I am interested in the ways in which DeLillo and other contemporary American writers have tried to find such a "refuge" for themselves, and to create a professional space where the writer can view events that do not, in and of themselves, have a novelistic coherence. DeLillo's fiction since *Libra*, in my view, realizes certain possibilities for contemporary literature to contest not only with the raw material of the real, the actual, but with the more pervasive and culturally dominant media that are all too quick to turn this material into *story*. DeLillo offers his own narrative as "a way of thinking about the assassination" that is different from, though clearly dependent on, the accumulation of fact that frustrates a scientific analyst such as Branch ("Author's Note"). Like Walter Benjamin in the *Arcades Project*, DeLillo seeks a constructive principle—and a *style*—by which to release the expressive power within historical facts and fragments that themselves remain unconnected. Rather than passively accumulating as the product of documentary research, the facts must be made to come together in a mosaic or nonrandom montage, to form, in Benjamin's words, a significant "construction of life."[6]

DeLillo's approach requires a certain disciplinary fluidity, a degree of professional crossover, and a corruption of genres that answers more than Tom Wolfe's wish that the contemporary novelist would begin with a journalistic approach to American matter. Unlike Wolfe in *The Bonfire of the Vanities*, DeLillo in *Libra* is not simply carrying the results of his "research" into the conventional novel of social manners; nor is he overly interested, as is Wolfe, in satirizing particular instances of human foolishness or institutional corruption. To understand an event that in its continuing recalcitrance seems, to DeLillo no less than Branch, "an aberration in the heartland of the real" (*Libra* 15), the contemporary writer must find a narrative form that does not presume a stable reality.[7] In *Libra* DeLillo achieves this form, in part by blending the mode of documentary with imaginative fiction, reworking (if not wholly undermining) distinctions between fact and fancy, truth and fiction, objectivity and subjectivity, and so giving psychological fluidity to "outside" documentable details.

The cultural motivations to create such a form, of course, go beyond the assassination. DeLillo's career and formal preoccupations are those of the late modernist writer who is becoming ever more separated from the age's dominant energies, and who must live amid technological forces and corporate systems that have grown too complex for any single mind or imagination to comprehend. Making imaginative use of the images and data that such systems produce is one way for the artist to build himself back into a culture that tends to exclude him, aesthetically making a virtue of one's cognitive limitations: "Establish your right to the mystery," DeLillo advises in an essay on the assassination published five years before *Libra*, "document it; protect it" ("American Blood" 27). For if a total naturalistic or sociological representation of events is no longer possible, if reality, as more than one of DeLillo's elder contemporaries has pointed out, is no longer realistic, that per-

6. All references to Benjamin in this chapter are drawn from Susan Buck-Morss's *Dialectics of Seeing: Walter Benjamin and the Arcades Project*. Benjamin's project was to be a part-analytical, part-documentary work on the construction of industrial and leisure city spaces which, had it been completed, would have made a fitting critical counterpart to the contemporary narratives under discussion here.

7. In the essay "American Blood," DeLillo first identified the assassination as "a natural disaster in the heartland of the real, the comprehensible, the plausible" (23).

ception in itself is a basis for creating a recalcitrant, ambiguous, and self-consciously "mysterious" fiction.[8]

This emphasis on ambiguity and the desire to find "mystery" in the world is surely in line with modernist and romantic oppositions to science. But the particular form that DeLillo would arrive at in *Libra* is more precisely locatable in American literary history. John Limon rightly identifies the blending of genres, in Hawthorne, Poe, and Henry Adams no less than in the contemporary nonfiction novel, as one typical literary response to a culture silently dominated by science, technology, and the cult of information. In books on the space program by Wolfe and Mailer, for example, generic flexibility "is at least partly an attempt to live in a scientist's historical moment and keep free of that moment and that history at once" (Limon 188). The same can be said of Henry Adams's *Education*, which represents an earlier attempt "to blend speculation on science with autobiography, egolessness with ego . . . invisibility, in every case, with personality" (188). In Limon's self-proclaimed "disciplinary" history of American writing, however, the literary response is said to be "hopeless," doomed to fail because literature and science possess separate and irreconcilable histories: whereas science "kills its past" as the scientist solves old problems that are then no longer worked on, literature does not, and this deep incompatibility makes it impossible for literature to progress with science into the modern world, even to oppose science. The modern writer, immersed like Adams in a universe of scientific and technological forces that exclude him, is thus left with only contradiction, ambiguity, and paradoxical expression; whatever he "wants to say is double," caught within two incompatible histories (188).

The marginality of the writer, Limon argues, is thus in part the result of literature's admitted failure to keep up with the professionalization of the sciences, a failure that no amount of ego loss or romantic posturing can make up. But that is no reason why the modern writer and the professional scientist should be seen as running on two wholly unconnected historical tracks. Even when the writer's work is offered as

8. In his early story "The Man Who Studied Yoga," Mailer portrays Sam Slovoda as a novelist who no longer wants to write "a realistic novel, because reality is no longer realistic" (*Advertisements* 179). More famously, in an article of 1961, Philip Roth noted that American "actuality [was] continually outdoing fiction" (120). Like Mailer (and DeLillo after him), Roth went on in his own work to explore the complex interdependence of fact and fiction, especially in *The Counterlife* (1986) and *The Facts* (1988).

an explicit attack against the sciences, scientific affinities can be found in the cultural and material constraints on both literature and science, and these shared constraints tend to reveal themselves most directly in the compositional activity that organizes and shapes the material of narrative. As Richard Poirier has pointed out in his essay "Venerable Complications: Literature, Technology, People," literature in general "appropriates, exploits, recomposes, arranges" not only materials that are said to belong to everyday "life," but also the material of earlier forms and genres that are perceived to have "become reified and potentially deadening" (126). Certainly the novelist may feel constrained by such inherited forms, if not by the history within language itself and the numerous social "codes" and inert physical structures that endow human life with meaning. But the novel, more than any other literary medium, has always managed to transform and recombine its inherited materials so as to express a reality perceived as multiple, complex, and threatening to our humanity.

As Poirier shows, and Limon implicitly acknowledges when he notes that technology tends to "channel all of a novelist's style into matters of 'technique' " (188), there need not be a fundamental difference between literary and technological modes of production, not if we consider the way the "original" modern novel gets written. The novel uses older forms as a kind of raw material every bit as much as the scientist or the engineer uses raw material in nature. And in doing so the novel is compelled to acknowledge, ever more directly in contemporary writing that must deal with an ever greater accumulation of reified forms, that "its own operations are akin to exercises of the technological power which it writes against" (Poirier, "Venerable Complications" 126). This self-conscious concern with the power of style and technique, which is everywhere evident in DeLillo, rarely leads to the universalizing, progress-oriented rhetoric that still characterizes much (though not all) professionalized and professionalizing science. (If anything, novels and fictive histories by these writers manifest a fluidity of narrative form that is comparable to the conceptual innovations of more recent and belatedly professionalized sciences such as the sciences of "chaos" and "complexity."[9])

Nonfiction novels, fictive autobiographies, novelistic histories, and

9. In his bestselling book *Chaos*, James Gleick tells the story of early scientific research in chaos theory. The essays collected by Katherine Hayles in *Chaos and*

documentary fictions: hopeless or not, much of the most culturally ambitious writing of recent decades has been characterized by a corruption of genres, so that more and more we find the science-minded novelist attempting to write, as Joseph McElroy attests, "not single-mindedly or generically stories, except as they would interrupt, interleave, break diverse kinds of document" ("Midcourse Corrections" 10). In his novel *Plus* McElroy worked within the science fiction genre to create a stylistic enactment of physical and intellectual growth, producing not an allegory of cultural progress into some utopian, visionary, or otherwise futuristic world, but a "fiction science" that expresses or imaginatively embodies present ways of knowing. DeLillo's use of mathematics as the main structural metaphor for his novel *Ratner's Star* led similarly to experiments in language and genre: "I began to find things I didn't know I was looking for," he says. "Mathematics led to science fiction. Logic led to babbling. Language led to games. Games led to mathematics" (LeClair and McCaffery, *Interviews* 86). Such hybrid fictions are less distrustful of science and of the various nonliterary media than they are wary of the peculiarly literary distortions that the romantic imagination tends to work on reality. As I mentioned in the introduction to this book, such fictions seem to me to have much to offer, not as a way of reclaiming for literature the centrality and professional status of the sciences, and not as defensive assertions of literary autonomy, but as a way of working more forcefully and eclectically at the margins of cultural information.

DeLillo is an author for whom a certain cultural marginality has always been a condition of writing. The situation of the "outsider," DeLillo's preferred term for himself, is essential to his work's very mainstream subjects, just as in his life he seems to have felt a "need to insert himself into yet remain alien from American life" (LeClair, *In the Loop* 14). Throughout his career DeLillo has tried to make a highly personal novelistic vision count within an institutionalized and technologized world system. His ambition is to represent or restructure the field of power and information in which he exists, though he must of necessity remain outside the representation, powerless to enact the restructuring. His work, like Adams's *Education*, is thus marked by a feeling of separation that is attributable neither to bourgeois individu-

---

Order, and Hayles's own book *Chaos Bound*, explore what the emergence of such "nonlinear" sciences as chaos and complexity might mean for the study of literature.

alism nor to romantic alienation, a feeling that the mind of the artist is not containable within the large contemporary forces that form it.

This situation of the outsider, a defining feature of the technological sublime, is also characteristic of McElroy and of novelists such as Gaddis and Pynchon, whose cultural ambitions have been accompanied by a notable elusiveness, and who, when not absent entirely, have tried in their narratives to project themselves beyond the enclosed perceiving subject into fields of power and networks of information that fragment and disperse the self. I do not think that this elusiveness is simply an accident of personality or the result of some New Critical desire to let the work stand on its own merits (and not on what the author or the popular media might have to say about the work). Nor are these writers attempting to compete with scientists by adopting an objective stance toward their subject matter. Even Pynchon, whose absence is usually interpreted as a refusal to collaborate with either the media, the marketplace, or the critical professorate, is there somewhere constructing and resisting his presence *in the language*—which is why the obsessive concern with "Control" and technical manipulation in *Gravity's Rainbow* is so hard to separate from the author's self-conscious troping of his own rhetoric, narrative management, and metaphorical manipulations.

Gaddis, too, though among the most reticent of authors, has made a point of constructing the character of the writer *in* the work, and not just the man behind the work, the "human shambles that follows it around" (*The Recognitions* 96). Gaddis would appear to have taken objectivity and a naturalistic presentation as far as it can go in *JR*, a massive novel that is nearly all dialogue, whose very movement from scene to scene depends on telephones, video cameras, transportation systems, and the whole range of crisscrossing communications networks that make up the corporate world. The book's harshest critics have interpreted Gaddis's refusal to write conventional narrative as a perverse American response to Roland Barthes's call for the "death of the author," and it is true that Gaddis suppresses his own personality and "literate voice" to become a medium for other voices, a recorder of the multiple cultural quotations that for Barthes make up any literary text (LeClair, *Art of Excess* 104). Yet Gaddis remains very much alive in the formal arrangement of the narrative, inserting literary allusions and details from his own biography that would be recognized by the more well-read among his readers, and creating structural ironies in an unbroken, recursive narrative that at every point resists the uncontrolled capital accumulation, illusory historical "progress," and rapacious technological "development" that the book documents.

Far from being a capitulation to an ideal of objectivity or the professional detachment of the sciences, the disappearance of the author would seem in every case to be a disappearance *into* the materials of contemporary culture, into sources of information that can themselves be made to express the writer's private "passions, moods, sentiments, impressions" (Barthes 53). Even in so objective a novel as *Libra*, whose "cold, clear sentences," in the words of one reviewer, "feel not so much fixed to the page as suspended above it, like fields of force sent through space," we are very much aware of the author who has self-consciously shaped the sentences, and who feels himself being shaped by them (Cain 275). The absence of an author is a great presence in this novel, as Gaddis is present in *JR*, and as Norman Mailer is present in "the rhythmical shapings that he gives to the paragraphs" and sentences of *his* first great documentary fiction, *The Executioner's Song* (Edmundson 441). As public personalities no two writers could be more different than Mailer and DeLillo, but with the publication of *Libra* in 1988 DeLillo can be said to occupy a cultural position not altogether unlike the position Mailer occupied in the sixties. And the shift in authorial stance that makes *Song* more than strikingly similar to *Libra* and *JR* is worth considering, since this shift toward an apparent objectivity suggests possibilities and limitations common to all those writers who left an oppositional politics behind in the sixties and seventies, and who have since sought a ground for resistance in the emerging technoculture.

Certainly *Song* is the most direct precursor to *Libra* (as Truman Capote's *In Cold Blood* was a certain inspiration for Mailer) not only in its focus on an obscure figure made famous by an act of violence, but also in the relation of the author to a range of journalistic materials. Gaddis had built the narrative of *JR* largely from newspaper clippings, government documents, and corporate talk he had heard and executive speeches he had written during his years as a corporate writer. Similarly, Mailer in *The Executioner's Song* sought to immerse himself in the detritus and documented reality of American life—specifically the life beyond New York and the literary politics that both writers had treated in previous work. Also like Gaddis, in the process of writing the novel Mailer became a medium for the whole range of local and corporate voices in America, the "western" and "eastern voices" of *Song*'s two halves. Mailer's characteristically knowing, judgmental, often hectoring voice is absent, as is any hint of his involvement in the cutthroat negotiations for the rights to Gary Gilmore's "story," the "senseless" murder of two Provo, Utah, men, the conviction to death in the electric chair, and Gilmore's subsequent insistence on having the sentence carried out.

The impossibility of determining a definite cause or reason for Gilmore's actions made the process of narrativization that much more urgent. The mainstream media predictably brought the full range of sociopsychological narratives to bear on Gilmore—precisely the kinds of collective narrative that Mailer, in his journalism of the sixties, had tried to supplant with his own aggressively subjective voice and immediate sense of events. In *Song,* however, he chose a different strategy and attempted to use the media themselves and the wealth of material they were capable of providing to sustain an illusion that the story is being told by those who knew Gilmore. The author's function, as in *JR* and *Libra,* appears to be a purely technical gathering of information, a matter in *Song* of allowing each one of the novel's typographically separate paragraphs to emerge as if from the notecard entries Mailer and his main collaborator, the sensationalist reporter Larry Schiller, recorded over hundreds of interview hours. Yet—to repeat—such detachment and objectivity are very much illusory. Mailer has not eliminated himself from the narrative; he has only situated himself *inside* its material, and—as in *Armies*—he has subjected his romantic and existentialist aesthetic to the constraints of documentable reality.[10]

After his earlier self-advertisements, Mailer in *Song,* and even more noticeably in *Harlot's Ghost* (1991), has ceased romantically to set himself against the culture. With Gaddis and DeLillo, he has sought a resisting style that is integral rather than oppositional. Instead of absorbing reality within the warlike, all-consuming masculine Self, he now attempts, through the slow accretion of historical details, to match himself with a reality that remains outside him, and outside the limits of his own authorial control. He has moved, in other words, away from the egotistical sublime that absorbs all otherness into himself toward a less ambitious documentary realism, a mode that in its mellowed-out humanism and quiet reliance on the media culture itself perhaps

10. Mark Edmundson argues, contrary to what most of *Song*'s first readers thought, that Mailer did not "succeed in refining his prodigious ego out of the book" after running out of ways to advertise himself in his earlier work. Rather, the disembodied narrative of *Song* represents yet another form taken by Mailer's romantic aspirations and "Emersonian illusions about originality and self-invention" (435).

no longer corresponds to the category of the sublime at all but to the beautiful.

To be sure, the compulsion toward invisibility has not caused Mailer, DeLillo, or Gaddis to write without ego, any more than Henry Adams's sense of his own marginalization kept him from writing autobiographically. But the desire for continued self-creation, real and persistent though it may have been, now seeks support outside the self, according to a constructive or compositional function. For DeLillo, "working at sentences and rhythms" is not only a source of aesthetic satisfaction, it is a way for a writer, over the years, to "begin to know himself through his language. He sees someone or something reflected back at him from these constructions" (LeClair and McCaffery, *Interviews* 82). This is not a narcissistic reflection; the compositional activity does not seek to *replace* reality with a self-contained aesthetic order (as in Cleanth Brooks's well-wrought urn or Wallace Stevens's aesthetic jar that, in giving shape and definition to the sloppy Tennessee "wilderness," ends up taking "dominion everywhere"). Unlike these modernist compositions and mythic substitutions, the arrangement and release of cultural data into a narrative flow intricates the postmodern literary imagination in nonliterary material.

The novelist's cultural immersion and narrative ventriloquism thus represent neither a loss nor a romantic displacement of ego so much as they become a way of testing the reality of the writer's novelistic vision, even if that means, in *Song*, Mailer's identifying himself obliquely with a criminal protagonist. For DeLillo, too, the re-creation of the life of a historical figure was an opportunity to try out a particular style of consciousness, to present the speech patterns of various American types, and to see if a private vision that had been maturing over eight previous novels could be made to coexist with the factual world of recorded history. The figure of Lee Harvey Oswald existing alone on the margins of society, a hapless systems builder who wants to immerse himself in history, comes as naturally to DeLillo as the psychopath Gilmore came to Mailer. Each protagonist represents the antithetical American that both writers had been dealing with all along, Mailer's "white Negro" and DeLillo's suburban terrorist, the would-be underworld hero and sometime stand-in for the artist. With such characterizations DeLillo and Mailer do not glorify their subjects so much as they generalize them, and so embed the artist's own metaphors, fictions, and imaginative structures in a plausible historical circumstance.

The inclusion of historical figures in fiction has become a fashionable convention of realism since the eighties, a way, it would seem, of poaching on the power of the more popular nonliterary media. Having abandoned fiction's truth claims, the contemporary novelist occupies the void left by the media, filling gaps in the historical record, and using the ready-made drama of a much-publicized yet unexplained act of violence to bring out the unformulated themes, hidden designs, and "secret symmetries in a nondescript life" (*Libra* 78). The degree to which such novelistic reconstructions depend on contemporary technologies, especially the media technologies that DeLillo and Mailer claim to write *against*, is evident in the smallest narrative detail—DeLillo's knowing what Oswald and his mother, Marguerite, might have viewed "Thursday nights" on television (*Libra* 5), or Mailer's reconstruction of Gilmore's dazed wanderings from place to place on the outskirts of Provo on the night of the murders. Not one of these narratives would have been possible except in a technological culture that preserves such information in library archives, information that constitutes, in Benjamin's phrase, a source of "ready language" that only waits on the writer to be put to narrative use (Buck-Morss 17).

The novelistic use of this ready-made archival language is not restricted, of course, to documentary fictions such as *Libra* and *Song*. Pynchon depended no less on London newspapers of 1944 for details of daily life during the rocket Blitz: the weekly airtime of the "Radio Doctor asking, What Are Piles?" (6:45 P.M., according to Steven Weisenburger); the recycling of Spam tins for children's Christmas toys and old toothpaste tubes "for solder, for plate, alloyed for castings, bearings, gasketry"; the precise time of an early rocket strike ("6:43:16 British Double Summer Time") which, according to reports, was widely mistaken for an exploded gas main (*Gravity's Rainbow* 133, 132, 26). Fragmentary and incomplete in themselves, such journalistic details often contribute little more than a fragment overheard by a passerby:

> Not a buzzbomb [Slothrop thinks], not that Luftwaffe. "Not thunder either," he puzzled, out loud.
>
> "Some bloody gas main," a lady with a lunchbox, puffy-eyed from the day, elbowing him in the back as she passed.
>
> "No it's the *Germans*," her friend with rolled blonde fringes under a checked kerchief. [. . .] (25)

Cumulatively, however, the mosaic of local fragments that constitute a novel such as *Gravity's Rainbow* can be made to suggest an entire experiential world. Rescuing such details from the objectifying history of the camera or the tape machine on the one hand, and from the pseudonarrational techniques of contemporary journalism on the other, the writer endows these details with a novelistic coherence, capable of speaking to human concerns.

Tom Wolfe has lamented the supposed failure of his more literary contemporaries to represent the effects of large social pressures on the individual ("Stalking the Billion-Footed Beast"). Such a representation, however, requires much more than the passive "research" that Wolfe sets as the task for the contemporary novelist. Writers of naturalistic fiction in America have traditionally invoked particular scientific or psychological mechanisms—Dreiser's Darwinism, Norris's thermodynamics, Dos Passos's camera-eye technology—to organize the welter of contemporary phenomena. And it is often noted that postmodern naturalists have been all too willing to let scientific forces decharacterize the people in their fiction, that these novelists make their characters subservient to abstract structures and connective patterns. Yet the contemporary writer does not share the traditional naturalist's naive belief in scientific causality, despite this continuing interest in the generalizing power of science. I have been arguing that the novelist comes to share most deeply in the technological culture by instead being receptive to the expressive power in its products and so bringing these otherwise mute forms into the realm of language, symbol, and metaphor. The writers I have been discussing do not depend for their own truths on an appropriated truth from science, any more than they rely on the "objective" truth of documentary history. They *construct* a truth by actively perceiving a narrative form in material that is real but not in itself linguistic.

Thus, DeLillo in *Libra* does not seek the "literal truth" of Oswald's role in the assassination ("Author's Note"); nor does he attempt an encyclopedic presentation of all that is known about Oswald. Rather, he plays off the known facts of the assassination against all that we do not know, and by inhabiting the space of our uncertainty he creates a fiction capable of inserting itself into the world. By establishing themes and connective patterns among the confusion of false names, blank spaces, and contradictory evidence that continues to shape speculations about Oswald and the assassination (in the Warren Report and academic histories no less than the popular media), DeLillo would provide us with a way of thinking about the assassination more intriguing than "the facts as we know them," even if he would produce a history that is, "in the end, a story about our uncertain grip on the world" ("American Blood" 28).

When *Libra* first appeared, it predictably enraged reviewers of the "media political right," whose aesthetically naive but politically serious attacks on DeLillo are defended by Lentricchia ("Citizen" 241). George Will called the book an act of "literary vandalism and bad citizenship," and Jonathan Yardley found DeLillo's use of still living historical figures such as Oswald's wife, Marina, to be "beneath contempt." Rather more substantial was Michael Wood's essay in the *New York Review of Books*, which pronounced DeLillo to be the successor to Mailer and Pynchon as the new "high priest" of literary paranoia in America (as if DeLillo had not delineated the debilitating effect of paranoia and conspiracy thinking *in* his characters). But for all its inherent mystery and complexity, the characterization of Oswald in *Libra* is indicative not of paranoia but of a remarkable generalizing power on DeLillo's part, a suspicion of orders beyond the visible, and a sense of immanent literary meaning in objects and constructed spaces that are themselves mute.

This wholly immanent "world inside the world," in DeLillo as in Pynchon and Mailer, is only incidentally linked to the particularized underworld of global politics and professional espionage (*Libra* 13 passim). As DeLillo had written in "American Blood," the mystery surrounding Oswald could be "simply what happens when we expose the most ordinary life to relentless scrutiny, follow each friend, relative, acquaintance into his own roomful of shadows, keep following, keep connecting. . . . We may all lead more interesting lives than we think" (24). Like any American, Oswald while he lived was immersed at every moment in relations of political power, whether or not he acted as agent or dupe of some larger conspiracy to assassinate the president. (The theory DeLillo ends up working with seems plausible enough to me: he presents Oswald as "a man with links to intelligence agencies but not necessarily guided by them, not duped by them, a man more childlike and lost than most theorists today will concede" ["American Blood" 28]. It is the downfall of the powerless to assume that those in power possess conspiratorial acumen equal or superior to their own: the most difficult premise to grant is the possibility of plain stupidity in the powerful, against which—in Goethe's words—even the gods fight in vain.)

In DeLillo's fiction the objective truth or falsity of such speculations remains beside the point, though DeLillo, unlike the popular filmmaker Oliver Stone, is careful to get his facts right. Stone's controversial movie *JFK,* which appeared a few years after *Libra* in 1991, merely replaced the selective and unlikely story of official culture with a slightly less dubious but equally one-dimensional story from the liberal left. (This is why Stone's conspiracy theory could be discounted in toto by the news media when contradictory evidence was brought forward by the physicians who conducted Kennedy's autopsy, even though such evidence was incidental to Stone's main contention that Oswald could not have been the only assassin.) By contrast, neither DeLillo in *Libra* nor Mailer in *Harlot's Ghost* has felt compelled to construct an unambiguous conspiracy around the assassination of Kennedy. DeLillo read the twenty-six volumes of the Warren Report not primarily for what they tell us about the assassination, but for the "extraordinary window" they provide "on life in the fifties and sixties" (DeCurtis, interview 292). The fruit of DeLillo's research is not a new theory of the assassination, but a more difficult theory of the intersection of large historical and technological forces with ordinary lives, "the daily give-and-take that's like a million little wars" (*Libra* 113).

DeLillo understands very well how readily conspiracy politics can become an escape from these less obviously dramatic struggles, and he knows, too, that the widespread fascination with conspiracy (even when justified by suspect omissions and apparent fabrications in the historical record) can be an expression of a culture's powerlessness, its distaste for everyday political life and impatience with the ordinary, the "daily jostle," everything that remains undramatized and unformed:

> If we are on the outside, we assume a conspiracy is the perfect working of a scheme. Silent nameless men with unadorned hearts. A conspiracy is everything that ordinary life is not. It's the inside game, cold, sure, undistracted, forever closed off to us. We are the flawed ones, the innocents, trying to make some rough sense of the daily jostle. Conspirators have a logic and a daring beyond our reach. All conspiracies are the same taut story of men who find coherence in some criminal act. (*Libra* 440)

In *Mao II* DeLillo would admit that the novelist, too, is prey to this feeling of powerlessness. The protagonist of that novel, the reclusive writer Bill Gray, feels that his work has been supplanted by the "taut" stories of terrorists and conspirators. He says: "What terrorists gain, novelists lose. The degree to which they influence mass consciousness is the extent of our decline as shapers of sensibility and thought. The danger they represent equals our own failure to be dangerous" (157). Yet the very precision of Gray's syntax, the exactitude and balance of the three parallel sentences equating novelist and terrorist, betrays a concern with order that is basically at odds with the disorder and imperfect control that any real conspirator would have to live with every day.

Where DeLillo in *Libra* takes his greatest risk and demonstrates his cultural power (I mean as a writer) is in locating much of the book's narrative in the solitude of Oswald's mind, and in the comparable minds of invented conspirators and various other men (always they are men) "at the mercy of [their] own detachment" (*Libra* 18). *Libra* opens with Oswald riding a New York City subway, a figure of the single male consciousness being "propelled headlong" through some enclosed space that recurs in opening passages of several DeLillo novels (Molesworth 381).[11] Though still a teenager, Oswald is already cast as an outsider, wrapped in himself and doubting the reality of people on the platforms and in the train. He personally is "riding just to ride" (3, 13) and keeping apart from the utilitarian routine in order to comprehend the totality of the life around him. Here, and throughout the novel, he seems to be watching himself watch. He feels the thrill of being driven to "the edge of no-control" (3) by a motorman who he thinks might be "insane" (13), but there is nonetheless a kind of order and security within his private pocket of machinery and speed, where he can watch the city go by and read it like a code: "One forty-ninth, the Puerto Ricans. One twenty-fifth, the Negroes. At Forty-second Street . . . came the heaviest push of all, briefcases, shopping bags, school bags, blind people, pickpockets, drunks. . . . There was nothing important out there, in the broad afternoon, that he could not find in purer form in these tunnels beneath the streets" (4).

Oswald's impulse to impose an order on a world he cannot comprehend is surely one that any writer might share in trying to create a novelistic order within the superficial chaos of story elements. As in McElroy's *Lookout Cartridge* and Mailer's *American Dream* (not to mention Dos Passos's *Manhattan Transfer*), the streets of Manhattan and the

11. Most directly the subway ride recalls the comic initiation of young Billy Twillig into the mysteries of the New York City subway system. Billy's father believed that a man "should introduce his lone son to the idea that existence tends to be nourished from below, from the fear level, the plane of obsession, the starkest track of awareness" (*Ratner's Star* 4).

Bronx provide the classic grid of American reality, a "concrete abstract" of the entire city life that an individual consciousness moves through at every moment (McElroy, "Neighborhoods" 201). Yet the technology that is perceived, by DeLillo no less than Oswald, as a possible source of order and control also encloses the self and threatens at any moment to go *out* of control. The underground noise and darkness "was a secret and a power" (13) promising a total knowledge and foreshadowing later systems in the novel that are said to "collect and process . . . [a]ll the secret knowledge of the world" (77). Such homologous systems in DeLillo are never wholly reducible to a particular scientific theory or vaguely indeterminate "mystery"; they are rather human-made mechanisms that at once exceed our understanding and confine us to small spaces.

Oswald himself is only too aware of being "a zero in the system," but from within this marginal condition he is made to share the writer's need "to sense a structure that includes him, a definition clear enough to specify where he belongs" (357). Oswald's desire, as expressed in a letter to his brother that DeLillo uses as the epigraph to part one of the novel, is to merge with history, to join in "the struggle, where there is no borderline between one's own personal world, and the world in general." Throughout his life, however, Oswald is excluded not only from the great political struggles but from any sense of social or psychological belonging. His childhood with his mother, Marguerite, is "a dwindling history of moving to cheaper places" (5), and later, in adulthood, the "stunted rooms where he'd spent his life" (100) become the outward expression of his exclusion.

"The recurring motif in the book of men in small rooms refers to Oswald much more as an outsider than as a writer," DeLillo points out (DeCurtis, interview 290). Yet more than once we see Oswald trying to write, and his writing becomes the room he occupies as he runs against "limits everywhere. In every direction he came up against his own incompleteness. Cramped, fumbling, deficient. He knew things. It wasn't that he didn't know" (*Libra* 211). Oswald's literal difficulties with the language, a dyslexia that is nearly disabling, become another way for DeLillo to dramatize the general powerlessness of experience in a subject whose abilities, I am afraid, are not far from the present American norm. In contrast with DeLillo's own fluency, Oswald's hopeless efforts and "vocational interest," expressed on a visa questionnaire, *"to be a short story writer on contemporary American life"* might have been all too easily turned into satire or parody (134).

But DeLillo avoids satire by recognizing in Oswald his own desire to put a self together out of the materials of contemporary history; both the creative and created subject thus hope to "extend" their private fictions "into the world" (50).

We observe Oswald, for example, in the literal closet where he worked out the failed assassination of General Walker, a figure of some notoriety from the extreme right: "What a sense of destiny he had, locked in the miniature room, creating a design, a network of connections. It was a second existence, the private world floating out to three dimensions" (277). Throughout the novel Oswald attempts continually to construct himself, to put himself down on the page and give form to the chaos of reality. His small room helps him sustain a concentrated illusion of control, a dogged belief that he could overcome "the struggle and humiliation, the effort he had to exert to write a simple sentence" (211). When Oswald does at last get his man in his gun sights, Walker is seen, ironically enough, doing his income taxes for the year. He is yet another man in a small room writing: "He scratched out numbers, added up tax dollars." But his mind was on his own fantasy of political control, a ceremony of "flags waving in halls across the whole damn state, the draped bunting, the clear American voices calling out a song" (284). The assassin and his target are nicely balanced—and self-canceling. For all their contradictory politics, the rhythms of their thought and obsessions with destiny are identical: the destiny is that of the Libran, whose sign is the balances, the scales that "can go either way," independent of any one person's intentions.

The novel as a whole is just such a story of multiple attempts at self-construction competing and canceling one another out. For even as Oswald believes he is creating his own destiny, he is already being constructed by others, quite literally caught in the mechanisms of conspiracies he knows little or nothing about. The former CIA agent Win Everett in his basement at home, hunched over the worktable, cutting and pasting a set of false documents for Oswald, is only one of many characters who are busy constructing an identity for Oswald, one of the "men who believed history was in their care" (127). Everett sees Oswald as one element in "a fiction he'd been devising, a fiction living prematurely in the world" (179). He literally "scripts" Oswald (just as DeLillo had done in "American Blood") "out of doctored photos, tourist cards, change-of-address cards, mail-order forms, visa applications, altered signatures, pseudonyms" (24), anticipating with a professional's pleasure the appreciation that future historians would give to the personalized subtleties and touches of realism in the characterization. Such a constructed order, however, not only tends to objectify people, to distance and "frame" them in visual images, but also threatens to isolate the observer, leaving him apart from others, "at the mercy" of his own detachment, and spying on himself as if from a distance (18).

By the book's end, the novelistic parallels have proliferated to such an extent that the reader might well ask whether DeLillo himself suffers from a similar isolation, a separation from sources of power that can only repeat itself in the author's detachment from the world he constructs. Such detachment, as Lentricchia reminds us, is not the distance of an "omniscient" author or literary experimenter who displays human desires and judges them from the outside ("Critique" 448). DeLillo does not stand above his characters; on the contrary, he often appears as helpless before the ironies of the world as the people he writes about. The sense that things could go "either way" infects DeLillo's narrative as much as it defines the psychology of his main protagonists, whose passion for control is oddly canceled by the conviction that the course of events cannot be resisted. Up to the very moment of the assassination, DeLillo seems purposely to sustain a sense of aimlessness and a schizophrenic flatness in which all items have an equal, and equally affectless, significance. The meaninglessness can be rescued only by the external reality of the assassination; everything else on that day in Dallas is experienced as a kind of simulation, as if even those present were seeing the event through one or another of its many filmed versions.

Thus, Kennedy's reality is established, as he greets the crowd at the airport, through the media image that would have been familiar to most onlookers: "He looked like himself, like photographs, a helmsman squinting in the sea-glare, white teeth shining" (392). From this point on in the chapter the movie camera never stops running. Events approach their violent conclusion independently of the actors involved, and Oswald's view of things through the gun sight is no more real or determinate than any other view. After shooting twice, Oswald sees the wounded president below him in the momentarily stopped Cadillac. He observes Governor Connally's "startle reaction," a term he remembers "from gun magazines," without ever thinking to connect the reaction to anything Connally might actually be experiencing (398). As Oswald prepares to shoot a third time (not the shot that kills the president), he is still figuring the possibilities in his mind, still detached from the reality of his own involvement in the event:

> Okay, he fired early the first time, hitting the President below the head, near the neck area somewhere. It was a foolishness he could dismiss on a certain level. Okay, he missed the President with the second shot and hit Connally. But the car was still sitting there, barely moving. He saw the First Lady lean toward the President, who was slumped down now. A man stood applauding at the edge of the telescopic frame. (398)

The presentation of Oswald's thoughts at this moment is at once deeply disturbing and wildly funny, a good example of the "terrific comedy" that Lentricchia identifies as "DeLillo's mode" ("Citizen" 240). This tonal instability and Oswald's pathetic attempt at self-justification, when he is not sure that he even *wants* to kill Kennedy, are indicative of the way media simulations and a spirit of self-watching can enter the very thoughts of a character. DeLillo's presentation of an indeterminate subjectivity, "the whirl of time, the true life inside" an ordinary personality such as Oswald's, is surely a major postmodern characterization. But in terms of DeLillo's larger project to represent history, the characterization remains problematic: what history can there be in a culture whose every action is so drained of consequence that only death can escape the play of rationalization and endless interpretation?

Mailer had experienced a similar problem in *Song*, and it is again instructive to consider Mailer's earlier handling of a reality that, because it has been abstracted from immediate human concerns, can always be made to go "either way," depending on who is in power. Gilmore, like Oswald, lives a life of no clear direction, adrift in a culture that never seems real to him. Even when he is sentenced to death, he has to insist on getting the sentence carried out, asserting that he had taken the court's order seriously, and that now he expected the people of Utah to take him seriously by letting him die. The sublime moment in *Song* emerges as if from sheer American substance, and from the figure of Gilmore himself as he enacted, before a rapt media audience, the white Negro's imperative to "live with death" (*Advertisements* 339). Gilmore, the first person in the United States to be sentenced to execution in over a decade, was also the only person in modern memory to choose the firing squad over more "traditional" methods of delivery (such as lethal injection or the electric chair), and he inaugurated his own death with the much-quoted hortative, "Let's do it."

In its macho style, Gilmore's ethic looks back toward Hemingway and an existentialist necessity to internalize death, socially and individually, in order to feel our present realities. But the detailed and difficult reality in *Song* may have more in common with a postmodern, postexistentialist culture increasingly defined by media simulations, in which death is the only way to put a stop to the infinite play of signification, the language games and endless proliferation of roles in the social sphere.

This tendency toward what Jean Baudrillard has called a culture of the "simulacrum" accounts for the difficulty not only of Mailer and DeLillo but of contemporary writers in general to imagine either a fictive or an actual "plot" that does not have, as DeLillo writes in *Libra*, "a tendency . . . to move toward death" (221). Like Everett, the obsessed conspirator who once broke down under the objective scrutiny of a lie detector, DeLillo in his fiction "would go beyond yes and no. Tell them about the deathward-tending logic of a plot" (363). He would even seem to present an alternative to this logic in the closing chapter of *Libra*, in which Marguerite Oswald mourns the death of her son.

In a narrative dominated by male conspirators who hatch their plots in small rooms, Marguerite directs her appeal to the unseen, unlistening representative of American power: "Your honor, I cannot state the truth of this case with simple yes and no. I have to tell a story" (449). Her "small, fragmentary narratives" of Oswald's family life, full of "sincere pathos," are unassimilable to the grand narratives of the novel's conspirators, less concerned with "posterity and the problems of representation" (Hantke 157). The story she tells, however, is in essence the same story that DeLillo has been telling throughout the novel. Like DeLillo, Marguerite would arrange a life out of scattered materials that resist logic and sense; she makes her pathetic appeal from the outside, standing "here on this brokenhearted earth" (454), but her thoughts follow the same pattern as Oswald's, and her words reverberate in the same small room of history that she shares with all the male conspirators in the novel. Oswald's "world inside the world," for example, becomes "stories inside stories" in Marguerite's monologue (450). "I have suffered like my son. We have the same construction," she says (454). Her ramblings precisely echo—and thus only parody—the concentrated plot-making activity of the conspirators; instead of opening out to other voices, her dotty logic heightens the book's claustrophobia. Not until his next novel, *Mao II*, would DeLillo fully explore the power within other voices, and within other, less directed stories. But the exploration would first involve him in a disturbing, because intensely personal, reconsideration of his own condition as an outsider, one of the most thoroughgoing critiques in contemporary fiction of a merely *literary*, late modernist aesthetic.

The Death of the Author: *Mao II*

DeLillo, like his protagonist Bill Gray, has never been "an autobiographical novelist. You could not glean the makings of a life-shape by searching his work for clues" (*Mao II* 144). Yet clues of a different sort abound in *Mao II*, and they lead to what DeLillo only half ironically calls the "true life" of the novelist (136): his books, his body, his hair that collects in a typewriter, even the tedium and labor of composition, through which "his sap and marrow, his soul's sharp argument might be slapped across a random page, sentence by sentence" (144). With Bill, DeLillo is not simply adding one more type to a series of obsessional male characters in his novels; rather, he appears to be going to the autobiographical source of all such characters—the shaping imagination that would identify the world with language but that is constrained, nonetheless, to occupy an enclosed, marginalized cultural space.

After nine novels and the popular success of *White Noise* and *Libra*, DeLillo deserved to write a book about a novelist. And if his story of a missing writer indulged the preoccupations of a self-proclaimed literary "outsider," it was also certain to find a public resonance. When *Mao II* appeared, Salman Rushdie had been in hiding for just over a year, and reviewers were quick to remark on DeLillo's own activism, through PEN, in response to the Ayatollah Khomeini's warrant for Rushdie's death. At any other time DeLillo's fascination with terrorists and novelists, and their frequent equation in *Mao II*, would have been dismissed as the mere expression of literary paranoia, another private attempt to escape the male prison of history and ideology. As it is, Bill's *self-chosen* isolation is reminiscent less of Rushdie than of J. D. Salinger or Thomas Pynchon, whose long-awaited fourth novel, *Vineland*, also appeared within a year of *Mao II*. Pynchon has not himself entered the public world, despite inevitable rumors that he would surface with the new book. But he did write a rare endorsement for

Mao II,[12] which establishes DeLillo's right to the biography not of Pynchon in particular, but of the great single and disappearing author at the heart of contemporary writing in America.

Like Pynchon, Salinger, and, to a lesser extreme, William Gaddis, Joseph McElroy, and DeLillo himself, Bill is a well-known novelist who has "engineered" his absence from mainstream media and the culture industry (143). In a time when advanced criticism accepts the death of the author as a natural fact, Bill has "simulated" a death of his own (140). His absence, paradoxically, only enhances his reputation, and the cultural vacuum his novels might have occupied becomes filled with other stories of "his disappearance, his concealment, his retirement, his alleged change of identity, his rumored suicide, his return to work, his work-in-progress, his death, his rumored return" (31). Bill holds on to his current book, partly from reservations about its worth and partly from his secret belief "that the withheld work of art is the only eloquence left" (67). Like Harold Brodkey, still another missing writer whose critical reputation seemed assured as long as he did not publish his much-talked-about work in progress, Bill lets the new book "take on heat and light" in his absence. "This is how he renews his claim to wide attention. Book and writer are now inseparable" (68).

DeLillo has never tried to disguise his recycling of old themes. Rather, with every new novel he has needed to test their reality over again, to create continuities with the media culture, and to establish significant redundancies against its fluidity and forgetfulness. "Old books haunt the blood" (73), says Bill, and the most personal gestures in *Mao II* take the form of borrowings from the author's earlier work—an old quip from *White Noise* about a middle-aged parent's confidence in his insurance coverage, a writer's fetish for "stockings" (the word, not the garment [110]), the by now familiar meditations on Hitler's bunker (31–32), on nature giving way to a mechanically reproduced "aura" (44), and on the collective unconscious as the Esperanto of multinationals: "Mita, Midori, Kirin, Magno, Suntory" (23). The motif from *Libra* of men in small rooms is also carried into the newer book, only now it is specifically the mind of the artist that is poised against "the private ciphers, the systems of isolated craving" on the one hand (10), and a reactionary majority on the other who have become "immu-

12. "This novel's a beauty. DeLillo takes us on a breathtaking journey, beyond the official versions of our daily history, behind all easy assumptions about who we're supposed to be, with a vision as bold and a voice as eloquent and morally focused as any in American writing" (*Mao II* jacket copy).

nized against the language of self" (8). In Yankee Stadium 6,500 couples file past the Reverend Sun Myung Moon to be joined in marriage; they are, many of them, sons and daughters of the television viewers who were "spliced into the image" in DeLillo's first novel, *Americana* (15): "They know [Moon] at molecular level. He lives in them like chains of matter that determine who they are" (*Mao II* 6).

The crowd in Yankee Stadium is not the cult of "eerie-eyed children" that parents in the audience take it for, nor is DeLillo simply opposing the individual Western Mind and a depersonalized mass consciousness (9). What is truly frightening about DeLillo's imagining of crowds is that they have so much in common with the writer's resisting and articulate consciousness. Collectively the followers of Reverend Moon are no less aloof than the reclusive novelist: "Everything they said and did separated them from the misery jig going on out there" (13). Against the age's atomized, incoherent, and proliferating versions of the self, a sharper collective selfhood is being established in "the drama of mechanical routine played out with living figures" (7). Obsessive repetition provides a stay against history, mortality, and the contradiction of technology's fast obsolescence. In the spirit of literary modernism, "a time-honored event" such as marriage is made anew, and a fragmentary temporal experience is channeled into a spatial and timeless order (4).

Bill, of course, would prefer to channel his own life into a wholly textual reality, to disappear, like Pynchon, into his books. But this textual existence turns out to be no less mediated and reproducible than the narrative that insinuates itself, like molecular chains in a protein, into the minds and bodies of the people in a crowd. Their collective awareness of the spectacle they are making produces the same "heat and light" as Bill's withheld novel (68).[13] Escape to an autonomous textual sphere is no longer an option when the writer, made "harmless" by the culture of consumerism, has no standpoint outside the media from which to resist its excesses: "The more books they publish," Bill complains, "the weaker [writers] become" (47). Even work that makes as much of an impact as Rushdie's *Satanic Verses* is destined to be overshadowed by the spectacle that one culture of fundamentalist belief has staged for another, more technologically advanced, culture. Can one in ten people on record as either attacking or defending Rushdie

13. DeLillo uses the words "heat and light" in *Libra* to describe both the crowd's self-awareness in Dealey Plaza on the day of President Kennedy's assassination (394) and the massive "ignitable" paper record of the event (14).

actually have read his novel? "In the West," says Bill, "we become famous effigies as our books lose the power to shape and influence" (41). His preferred position as a cultural outsider, and his ambition to shape contemporary consciousness, have been occupied and taken up by "bomb-makers and gunmen. . . . They make raids on human consciousness. What writers used to do before we were all incorporated" (41).

DeLillo, like other "paranoid" writers in America, has often been criticized for naively confusing his own distance from sources of contemporary power with a general cultural malaise.[14] I wish to show here that the writer's private struggle with language, instead of separating him from power, is very much at one with narratives of power in the world, not excluding those operations of technological power that appropriate, exploit, and gather people into the material of a contemporary "history." This affinity between technology and literary production certainly creates dangers and complicities for the novelist, and *Mao II* is about all the ways in which the mainstream writer in America can become absorbed and incorporated not only as a "consumer event" (43) but as yet another cultural narrative, a feature in the constructed history of a nation, and the occasion, as we shall see, for an increasingly institutionalized criticism to practice the formalisms and methods that justify *its* autonomy.

In the wake of literary modernism, cultural technicians of various sorts have been only too pleased to regard the writer, and the writer's created world, as continuous, impersonal collections of signs having no fixed historical reference. The presumed death of the author consists essentially in this passage—or, more accurately, this deeply mourned pass*ing*—from a monolithic literary existence into a much wider range of "texts," for which no single creative consciousness could be uniquely responsible. To extend Roland Barthes's well-known formulation, the author's death has served as the condition of the birth not only of the reader but of the photographer, the social scientist, the terrorist, and the literary critic (55).

A late modernist writer who can no longer pretend to escape from

14. Interestingly, one of the best defenses of DeLillo's American "paranoia" comes from overseas. In a *Times Literary Supplement* review of *Mao II*, Julian Loose acknowledges the naïveté of "Bill's notion that writers were ever in a position to 'shape and influence' society." But Loose adds that "Bill's predicament dramatizes the difficulty facing any writer who, like DeLillo, is ambitious enough to want to make an impact beyond the literary" (20).

the mass culture into an autonomous sphere, Bill must learn to discover the enabling conditions of a literary existence from within the crowds and their defining media of communication. He decides, quite simply, to have a series of photographs taken, "to break down the monolith" he has built of his life (44). He chooses a photographer, Brita Nilsson, a woman in many ways like himself (and like a blonde Laurie Anderson in her looks and outlook) who has built an odd career photographing writers, preferring those "who remain obscure" (25). In her work Brita tries to be self-effacing; she "eliminate[s] technique and personal style to the degree that this is possible" (26), modestly claiming that her pictures have only a documentary value. She welcomes Bill's offer to pose for her as "a form of validation, a rosy endorsement" of a marginal obsession (66). But as the session gets under way, we soon understand that *she* is the more active and culturally central artist, for whom space "clos[es] in" during the course of a good working day (38). Bill himself had felt this way once, but now he spends many of his writing days "staring past the keys. Used to be that time rushed down on him when he started a book, time fell and pressed, then lifted when he finished. Now it wasn't lifting" (54). At the start of the session, Brita had felt "the uneasy force, the strangeness of seeing a man who had lived in her mind for years as words alone—the force of a body in a room," but soon she senses that Bill is "disappearing from the room," merging into a photographic image as he had once hoped to merge with his book (35).

Through just such symmetries, repetitions, and juxtaposed obsessions, DeLillo links the writer's own forms with homologous nonliterary forms, even if, for Brita, the writer would appear to remain separate and aloof. A sympathetic and worldly reader, she feels herself nevertheless to be an "outsider, not able to converse in the private language" of the writer (37). Bill sensibly counters that "the only private language" he knows "is self-exaggeration" (37). But even as he speaks and flirts with the photographer, we may sense another private language in which Bill is invoking, or being made by DeLillo to invoke, all those novelists of the sixties and seventies who may have served as models for DeLillo's own work and career. At one moment Bill's words recall Pynchon, whose failure to surface has been taken, by those who adhere to certain romantic notions of a writer's originality, as "a local symptom of God's famous reluctance to appear" (36). A moment later Bill recalls Norman Mailer in his approach to middle age, despairing of ever altering the "inner life of the culture" (41). And always there is the hunted figure of Salinger being surprised from behind by a photographer (37). The "human shambles" that Brita hopes to remake through the power of her seeing is very likely an allusion to Gaddis, who has long held that an artist, once his work is completed, becomes nothing more than a "shambles of apology."[15]

The private allusions do not end with the contemporary writers who have influenced DeLillo. He also inserts oblique references to the body of criticism that, like the critical industry surrounding Pynchon, has begun to grow around his own work. The night before the photographic session, Bill's assistant, Scott, had given Brita a tour of the house, in which every imaginable document having to do with Bill has been carefully stored and catalogued. The tour soon takes on the quality of a psychological descent into the living mind of the writer:

> They carried their wineglasses out along the hall where there were shelves filled with booklength studies of Bill's work and of work about his work. Scott pointed out special issues of a number of quarterlies, devoted solely to Bill. They went into another small room. . . . They went to the basement, where Bill's work-in-progress was stored in hard black binders, each marked with a code number and a date for fairly easy retrieval and all set on freestanding shelves against the concrete walls. . . . She waited for Scott to call this room the Bunker. He never did. And no hint of ironic inflection anywhere in his comments. But she sensed his pride of stewardship easily enough, the satisfaction he took in being part of this epic preservation, the neatly amassed evidence of driven art. This was the holy place, the inner book, long rows of typewriter bond buried in a cellar in the bleak hills. (31–32)

Scott's lack of irony would be unimaginable in any period of literary history other than the present, when entire journals such as *Pynchon Notes* are devoted to a living writer, and when new work, if not ignored entirely, is often absorbed in the same scholarly and archival spirit as the work of past writers. In a sense this basement room is the fulfillment of Bill's (and DeLillo's) worst hopes for himself: the writer has built an existence wholly in his books, and for this he must watch himself become the material of other people's interpretations. Bill senses the young critic at his brainstem "like a surgeon with a bright knife"

15. Cf. *Mao II* (37) and Gaddis's 1955 novel *The Recognitions* (96). Throughout *Mao II* Bill thinks of his unfinished book as an invalid and "hated adversary" (55), after a similar conceit in Gaddis's second novel, *JR* (603–4).

(38). Scott has learned to be a better reader of Bill's work than Bill is himself; and although he always maintains a respectful distance, his discretion only confirms the critic's professional authority to speak for the author. (When Scott summarizes Bill's thought for Brita over dinner, he speaks to her only, as though Bill himself were not in the room.) Scott says he is "only trying to secure [Bill's] rightful place" (73), but he has succeeded mainly in keeping the writer "in the room, seated" (64), while he himself has come into possession of the living text. Even after Bill abandons the room, Scott goes on making lists in his absence, organizing reader mail, cataloguing the photographs taken by Brita, and letting the unpublished novel go on "collecting aura and force, deepening old Bill's legend, undyingly" (224).

With Bill gone, the last words in the novel proper (before the Beirut epilogue) belong to Scott, although he is only, as always, "quoting Bill" (224; cf. 72, 119). The author thus passes into a critical technology of reference, a protocol for connecting signs to things that is independent now of the author's intentions or the connections the author may have built into the work: "What we have in front of us represents one thing. How we analyze and describe and codify it is something else completely" (222). In *Mao II* the writer's life is constantly being made into somebody else's material; his work is forever being mined and developed, documented, updated, and revised for consumption in ways that are parallel to the act of literary composition but that are never subject to a wholly literary control: "There's a force," says Bill, "that's totally independent of my conscious choices. And it's an angry and grudging force" (45). Never wholly master of the many social, political, and aesthetic contexts in which the work will function, the author belongs to a field of literary and cultural quotations very much like those that DeLillo has embedded in *Mao II*. Indeed, DeLillo's art in this book is mimetic not of the social and political order directly, but of the technological and literary-critical processes through which a writer might enter the order.

This welter of quotations does much to deepen the novel's private and oddly familiar character, and suggests that DeLillo is indeed working in a narrowly defined, self-admittedly "marginal," yet extraordinarily well documented tradition of innovative writing in America. In particular, one likely private reference to Tom LeClair reveals DeLillo's acute consciousness of the cultural field he is about to enter. LeClair had introduced DeLillo, Salinger, Gaddis, and Pynchon to general readers in an essay in *Horizon Magazine* (October 1981) titled "Missing

Writers," and it was by "reading a magazine piece about Lost Writers" that Scott was first put on Bill's trail (*Mao II* 58). LeClair is not, like Scott, the caretaker of the novelist's reputation, but he has done as much as any critic to establish DeLillo's position as a literary heir to the line of Herman Melville, Henry Adams, the turn-of-the-century American naturalists, and later "artists of excess" who have sought a literary equivalent to the range of American systems and structures.

Many critics besides LeClair have pointed out the extent to which these writers have all depended on the institutions of patriarchal and technological power they often seek to oppose, and DeLillo himself has never tried simply to escape such power. Nor has he tried, like Bill, to escape the critical professorate that can oppress even the authors it generously includes by making them the material of an objectified literary history. On the contrary, with this novel DeLillo has taken on a greater visibility and a more active public presence than ever before, and when he alludes to his own critics and peers, he would seem to be interested not in consolidating a place within a static literary "canon," but in bringing its energies more actively to bear on the wider culture.[16]

Like LeClair and the authors LeClair justly celebrates, DeLillo refuses to let key contemporary terms such as *excess*, *systems*, and *mastery* remain wholly the property of a hegemonic corporate power. His systems are not ruling systems, and a novelistic mastery does not simply reproduce or legitimate nonliterary technologies of organization and control. Like any writer, DeLillo needs to give himself to the narrow structure of a tradition, but he also needs to keep the tradition open to private and political languages that tend to escape compositional control: the language of crowds and street people, "the rag-speak of shopping carts and plastic bags" (180), a television "picture of people massing in a square" (185), hair in a writer's typewriter representing "everything that is not clear and sharp and bright" (201). Such languages are the necessary "excess" that enable the tradition—a literary order and cultural field of quotations—to replicate itself, to establish

16. Around the time *Mao II* was published, the *South Atlantic Quarterly* devoted an entire issue to DeLillo's fiction (Spring 1990), and DeLillo allowed an interview and photograph to appear in the *New York Times Magazine* (May 19, 1991). He also spoke with Maria Nadotti for a popular Italian literary journal, *Linea d'Ombra;* the discussion appears in translation in *Salmagundi* 100 (Fall 1993). In August 1992, however, he discouraged David Foster Wallace from going ahead with his plan of guest-editing a DeLillo number of the *Review of Contemporary Fiction*, and in July 1991 he had turned down an invitation from Dennis Barone to contribute to a *Review of Contemporary Fiction* special issue on Paul Auster.

feedback loops with the nonliterary world, and thus to unfold into new orders that have not yet been discovered or created. Rather than leading to a self-enclosed aestheticism, LeClair's, DeLillo's, and related narrative systems lead outside the self, into an open field of analogical relationships that dramatize the inability of any one mind to control reality.

In *Mao II*, I think more than in any of his previous novels, DeLillo purposely extends himself into areas of the culture that *cannot* be mastered by literary means. Even in mass marriage he recognizes a cultural impulse (uncritical and subject to manipulation though it obviously is) to "survive as a community instead of individuals trying to master every complex force" (89). In contemporary phenomena of such proportions, Bill's obsessive systematizing, his haughty distance and dogged need to "exceed" limits (39), may be less valuable than the experience of characters such as Brita and Karen, the latter a onetime Moon disciple who was among the couples in Yankee Stadium. (She does not necessarily like hearing the word "Moonie," even after she has been "deprogrammed" and has come to live, for a time, with Scott and Bill [77].) Bill may well represent a high modernist dead end for DeLillo, but Brita and Karen mark a continuing vitality in his writing—not only because they are among his more fully drawn and sympathetic women, but also because they take the narrative out of Bill's room and into the realm of competing voices, mixed populations, and nonhegemonic, even "terrorist," institutions.

Karen, who is able to anticipate the remarks of television commentators and mimic their "voices with a trueness that's startling" (66), becomes a primary vehicle through which nonliterary languages get processed in the novel: "She was thin-boundaried. She took it all in, she believed it all, pain, ecstasy, dog food, all the seraphic matter, the baby bliss that falls from the air" (119). Bill, sounding this time like William S. Burroughs on language as a virus from outer space, says that she carries "the virus of the future." But her command of verbal language is shaky, and her viral presence is not at all alien or alienated. Like a virus, Karen alters and contaminates whatever environment she enters. She does not use language to hold the world at a distance.

It is from Karen's permeable viewpoint that DeLillo narrates the crowd scenes reproduced photographically before three of the novel's four sections: she is there somewhere among "the thousands, the columned mass" filing past Reverend Moon (10), and she easily projects herself into satellite television pictures of the mourners at Khomeini's wake (105) and of human bodies pressing against a fence at a soccer riot (17). The soccer photograph, writes DeLillo, "could be a fresco in a tourist church, it is composed and balanced and filled with people suffering . . . a crowded twisted vision of a rush to death as only a master of the age could paint it" (33–34). The mastery of such a scene is partly DeLillo's, yet it is equally the product of the anonymous corporate media that record and transmit the photograph from which DeLillo is working. The digital array of photographic information is matched by a verbal precision and density of observation, and the author approaches not the impersonality of an Eliot or a Joyce but the neutrality of electronic systems of information. The self-conscious reference to mastery in this scene suggests that, to an extent, both the writer *and* the media share a modernist obsession with imposing an aesthetic order on the world. Yet DeLillo does not, like the modernists, allow his art to rise above the noise of contemporary culture. Through Karen's semiliterate and nonjudgmental consciousness, he and his text press up against a chaotic reality, matching the self with the range of contemporary "voices" and "images," and constructing a convergent, rather than a transcendent, order.

DeLillo never ceases to be a compositional presence in his narratives; he is always there in the sentences, shaping and patterning the inarticulate, and building an immanent authorial presence even as he proposes, in high modernist fashion, to escape personality. Yet he differs from the modernists in his refusal to replace reality with a merely aesthetic, self-created order, which, like Stevens's jar in Tennessee, would take "dominion everywhere." However preoccupied he may be with his own precariously balanced fictions, he never simply *opposes* them to a chaotic social and political order. Like Samuel Beckett, the one literary artist who is cited by name in *Mao II* (157), DeLillo seeks a novelistic form that would be "of such a type that admits the chaos and does not try to say that the chaos is really something else" (Gontarski 245). DeLillo is only too aware of how cloying "rhythms and symmetries" can become when they are pursued for their own sake (DeCurtis, interview 294), and he has Brita abandon her early modernist experiments in photography because, as she tells Scott, "no matter what I shot, how much horror, reality, misery, ruined bodies, bloody faces, it was all so fucking pretty in the end" (*Mao II* 24–25).

Because so much in *Mao II* cannot be circumscribed by aesthetic solutions, DeLillo lets significance emerge indirectly. The multiple parallels, symmetries, and recursive patterns in the novel do not replace a disturbing reality with something else (a private psychology, mythic cultural totality, or some other sublime image of the writer's own mind and imagination). Rather, and I think this is crucial to DeLillo's postmodern aesthetic, they enable the author to find the places where language converges with the real, the unpresentable, everything that does not conform to formal pattern and syntax.

Bill begins to practice just such an aesthetics of convergence when he abandons the closed room of his book to work for a worldly cause, the liberation of a young Swiss poet who has been taken hostage by a small Maoist group in Beirut. At first Bill hardly thinks of the hostage; he is merely playing a role that will satisfy the cute symmetries in a public relations negotiation: "I want one missing writer to read the work of another," says Charley Everson, Bill's old friend and would-be editor. "I want the famous novelist to address the suffering of the unknown poet" (99).[17] But Bill's attempt simply to turn up at a press conference in London and read some poems is frustrated, for no clear reason, by a terrorist's bomb. Later, George Haddad, an intermediary for the terrorist group, secretly proposes an alternative plan, a meeting between Bill and the group leader, which rather touchingly appeals to Haddad's "sense of correspondence, of spiritual kinship. Two underground figures. Men of the same measure in a way" (156). The meeting, however, never comes off; no individual action of Bill's, however slickly produced or professionally thought out, can prevent the terrorist's arbitrary assertions of control.

Not until much later, when Bill begins writing about the hostage, does he come to understand the uselessness of both Haddad's and Everson's plans, and not until he projects himself from his hotel room into the basement room of the hostage does he perceive the deeper symmetries and necessary differences between his own and the terrorist's separate narratives:

> When you inflict punishment on someone who is not guilty, when you fill rooms with innocent victims, you begin to empty the world of meaning and erect a separate mental state, the mind consuming what's outside itself, replacing real things with plots and fictions. One fiction taking the world narrowly into itself, the other fiction pushing out toward the social order, trying to unfold into it. (200)

17. The *New York Times Book Review* (January 14, 1990) staged a similar textual encounter by assigning Salman Rushdie the review of *Vineland*, a message direct from the underground pronouncing the voluntary exile's return a "triumph."

The mind that alternately empties the world of meaning and consumes meanings outside itself repeats a familiar pattern in the discourse of the sublime—specifically, the alternation between metaphorical and metonymical poles in Weiskel's analytic. In the first, *metaphorical* mode, "the absence of determinate meaning becomes significant" and the mind "resolves the breakdown of discourse by substitution," in effect setting up a separate mental state of its own. "The other mode of the sublime may be called *metonymical.* Overwhelmed by meaning, the mind recovers by displacing its excess of signified into a dimension of contiguity which may be spatial or temporal" (Weiskel 28, 29). Twice we have observed the tendency, in the cases of Pynchon and McElroy, of the sublime imagination to experience the world as if it were language. Nowhere is the danger of this textualist mode of experiencing so clearly perceived as in the present passage in DeLillo.

In his initial attempts to "unfold into" the hostage's unimaginable reality, Bill feels that he may be on the way to recovering "the shattery tension, the thing he'd lost in the sand of his endless novel" (168). But this imaginative projection into the room of the hostage has been made once before in the novel by DeLillo (107–12), so that by now all three writers—Gray, the hostage, and DeLillo himself—have begun to merge. The way is open for a preeminent modernist solution, in which Bill surmounts his compositional difficulties and transcends the pointlessness of contemporary history to become the author of the book we are reading. But, of course, DeLillo's narrative does not resolve itself so neatly. The only transcendence in the novel is achieved by the "new culture, the system of world terror" that the hostage has "tumbled into." Only in this context is he ever given a name:

> They'd given him a second self, an immortality, the spirit of Jean-Claude Julien. He was a digital mosaic in the processing grid, lines of ghostly type on microfilm. They were putting him together, storing his data in starfish satellites, bouncing his image off the moon. He saw himself floating to the far shores of space, past his own death and back again. But he sensed they'd forgotten his body by now. He was lost in the wavebands, one more code for the computer mesh, for the memory of crimes too pointless to be solved.
>
> Who knew him now? (112)

The hostage's "second self" is a weightless cybernetic version of the "second self" that Bill has grown in his workroom (37), the self that will go on after the physical body ends in pain and obscurity. The enforced disappearance of a hostage is the political counteraesthetic to Bill's disappearance; both help to produce an autonomous sphere in which "real things" are replaced "with plots and fictions" (200). But this projection into a cybernetic virtual reality is not limited to the extreme cases of hostages and reclusive authors. Throughout *Mao II* people are shown living in airless, self-referring rooms that afford scarcely any time or space for thoughts not relating to the self. Immediately outside the hostage's room, in the streets and on the radio, people talk "about Beirut because there's no other subject." Brita, who goes to Beirut on a visit, "wants to stand inside it. It is wrapped all around her like some computerized wall of enhanced sensation" (238).

The digital matrix Bill imagines is also the cyberspace of world monetary exchange, and it is appropriate that the young poet, his identity reduced to a digital image, is literally exchanged by his captors for another hostage at the book's end. The terrorist power of DeLillo's narrative comes from those moments when human bodies erupt into the placeless, selfless sphere of electronic transcendence. It is the only ideal out of time by which DeLillo can measure our fallen state, in which the material world nonetheless continues to press in on the consciousness of every character in *Mao II*, not only the writers, and not only in Beirut. Brita sees "the dumbest details of her private thoughts on posters or billboards" in cities all over the world, and people she knows keep turning up in picture magazines and airplane movies (165). Everywhere reality closes in, as if the world itself were language, down to the "language of soot" that settles in the skin of a bag lady and becomes the "texture" of her person (180, 145). Like people in Beirut whose "only language is Beirut" (239), street people in New York City living in boxes and layered garments are "edged down, reading the space their lives are assigned" (194). Officially "homeless," they are neither exoticized by DeLillo nor excluded from his presentation of the wider life of the city. They have created for themselves "a world apart but powerfully here" (149), which is as much autonomy as DeLillo would claim for fiction.

DeLillo's novels have always resisted the impulse to transcend their own materiality, not only in words but in the human body, in manufactured objects, even in the printed circuits of metal and silicon that make possible the seemingly weightless communications of modern electronics. Promises to free humans from material encumbrances have always appeared innocuous enough, and technology's practical viewpoint has its bland literary spokesman in Haddad, the Maoist theoretician who "absolutely swear[s] by" a new word processor by Panasonic. "It's completely liberating. You don't deal with heavy settled artifacts. You transform freely, fling words back and forth" (164). This is sublime comic relief, the eerily harmless obsession of a man normally preoccupied with terrorist negotiations to gain the "maximum attention" for the cause (164).

DeLillo is no technophobe; he can let readers laugh at Bill's old-fashioned resistance to contemporary technologies of composition and textual processing. As much as any contemporary writer, he has allowed his own language to play against the various languages of modern technology, to the point that he will often seem to disappear into the anonymous media that process the documents, photographs, sounds, and sights of contemporary culture. But these multiple texts are never wholly taken lightly; DeLillo never loses sight of the embodied reality beneath the information grid. He works with mediated representations not to deny reference but to make us see a hooded hostage's "face and hands in words" (160), make us "believe there is actually something under the hood" (205).

Epilogue

Postmodern Mergers, Cyberpunk Fictions

> In an amazing acceleration, the generations precipitate themselves.
>
> Jean-François Lyotard

More explicitly than any other work treated in this study so far, *Mao II* presents the late modernist writer at the moment of his passing from a monolithic existence in text into multiple technologies of reference. Hence, DeLillo's novel may stand as a fitting, if somewhat downbeat, conclusion to a study of a literary aesthetic that would integrate (without futilely replicating) its own forms and contemporary forms of technological production. With the demise of Bill Gray and the electronic transcendence of the hostage Jean-Claude Julien, we reach a troubling moment in the development of contemporary writing, a crisis of reference and representation that has pushed mainstream experimental fiction toward ever more weightless, disembodied, and self-consciously textual spheres.

Each of the novelists considered so far has wrestled with the scientific, the historical, or at least the journalistic "facts" that are capable of being tested in the outside world. But even these "facts" ultimately take a textual form, and *Mao II* so resists the contemporary tendency to represent the world and author as wholly textual that it stretches against the limits of the literary medium itself. In the very look of the novel, we may sense a reaction against textuality, a determination to test the limits of a matching language that continually brings identity back to its materiality in any number of nondiscursive media. DeLillo's inclusion, for example, of digital photographs that the text only comments on, together with a relentless self-consciousness about all that is "lost to syntax and other arrangement" (*Libra* 181), points to the primacy of sound and the visual image. For all DeLillo's verbal virtuosity (itself a part of the technological culture that insists always on the best possible performance), reality emerges from within DeLillo's textual figurations only as an immanent, nonverbal "mystery."

My concentration in this closing section will continue to be on fiction, but I should at least acknowledge that these very issues of referentiality and literary representation have been taken up, often with an "embarrassing literalness,"[1] by those who are working at the recognized "cutting edge" of practical technology. Advances throughout the eighties and nineties in "virtual reality" engineering, in "expert systems" of corporate management, and especially in "hypertext" technologies which integrate verbal, audio, and visual images have all helped to resituate literary texts within the greater field of nondiscursive media. Even readers who are not aware of contemporary theories of the world as text have had to imagine new relations "between a book's component parts, between the book and all the non-discursive apparatuses that surround it, enable it, and are transformed by it" (Brande 193). Indeed, the popular perception of a new textual relativity has been dramatic enough to have been noticed by *Time* magazine and the *New York Times Book Review,* in polemics that not only sustain perennial humanist concerns with cultural literacy, but also revive actual literary debates from the sixties about the death of the novel, the death of the author, and the postmodern literature of exhaustion (debates that were themselves largely inspired by McLuhanesque experiments in "multimedia" technologies). Robert Coover, an older contemporary of DeLillo and veteran of such debates, has gone so far as to suggest that current technologies herald an "end of books." This is perhaps a conclusion that I should not pursue at the moment, in this medium, although in closing I wish briefly to consider the referential field that I believe underlies the apparent desire of contemporary fiction after Mailer, Pynchon, McElroy, and DeLillo to escape its own materiality.

1. The phrase is George Landow's in *Hypertext* and in the collection, which Landow co-edited with Paul DeLany, *Hypermedia and Literary Studies* (6). William Gibson also complains that "some entrepreneurs seemed to have missed a certain level of irony in his writing," for Gibson himself did not anticipate the corporate push to turn concepts such as cyberspace and virtual reality into actual marketable products (BBC-TV, *Horizon* program on virtual reality, October 1991; reported in Christie 182).

By now it should be clear that the fiction analyzed in the foregoing chapters—representing a first-generation technological aesthetic that originated when computer and multimedia technologies were still new—has struggled against disembodiment and the culture of the image by various means, although its resistance may be, as Piotr Siemion has said of William Gaddis's *JR*, "ultimately quixotic." As we have advanced from a post–World War II period of expanding technological potential to a period of practical consequences, we are witnessing material actualizations even vaster, though probably less exciting, than those imagined in literature. The corporate and political world itself, says Siemion, has had "no place or use for infinitely accurate, infinitely demanding, non-traditional communications" of the sort considered here (private correspondence). Mainstream authors who have continued to work the technological terrain staked out in the sixties and not mined to exhaustion in the seventies and early eighties have increasingly bypassed the details of technological operations in favor of their simulations. Instead of the conceptual naturalism and experimental exuberance such details inspired in Pynchon, McElroy, DeLillo, Coover, and Gaddis, we are now offered, in Siemion's words, "the synthetic, simulated vistas of William Gibson's *Neuromancer*."

The so-called cyberpunk fiction of the eighties and nineties clearly owes a debt to postmodernist literary experiments of the preceding decades. Raymond Federman lists McElroy's *Plus*, Pynchon's *Gravity's Rainbow*, and DeLillo's *Ratner's Star* as obvious precursors, along with works such as Samuel Beckett's *Lost Ones*, Russell Hoban's *Riddley Walker*, Federman's own *Twofold Vibration*, and Italo Calvino's *Cosmicomics* and *T/Zero*, which have not been discussed here, but which might be included among those self-conscious, often cerebral tours de force that posit fiction as primarily a growth of language. Without these books, notes Federman, "there wouldn't be any Cyberpunk Fiction" ("Forum" 37). Federman, however, wrongly accuses his younger peers of failing to give credit where credit is due, and of much else besides, evincing a resentment that is, I would guess, the surprise of a pure postmodern experimentalist and programmatic antirealist at the practical realization of an a priori project. The difference between cyberpunk and its postmodern textualist predecessors is the difference, in the aesthetic sphere, between paper science and lab science, ideal and practical reason, the material advance of current computer technologies and the previous generation's heady theory.

Gibson himself takes the edge off such discussions by blithely acknowledging his literary antecedents, but without according them any greater influence than that of artists working in other media. He says, "I've been influenced by Lou Reed, for instance, as much as I've been by any 'fiction' writer" (interview in McCaffery, *Storming* 265). Even more encompassing is Bruce Sterling's announcement, in his definitive preface to the *Mirrorshades* collection, of "a new kind of integration. The overlapping of worlds that were formerly separate: the realm of high tech and the modern pop underground" (345). Federman had accused cyberpunk of not being explicit about just *what* was being newly integrated ("Forum" 37), but the consequences of the previous generation's experimentation reveal themselves quite clearly in this quotation: in an era of literary exhaustion and market saturation, possibilities might still expand and growth continue—in aesthetics no less than in economics—through hybridization, the corporate merger, the fluidity of seemingly inexhaustible connections across media and genres that are themselves perceived as defunct.

The cyberpunk writer thus typically sees himself (and until work by Pat Cadigan and Marge Piercy, cyberpunk had been, like its postmodern precursors in technology and literature, mostly a male genre) as leveling distinctions between the technical and the literary, fiction and history, "high" and "popular" cultures. This generic flexibility, while generally seen to characterize cyberpunk's postmodernism, is also a feature of the sublime moment in both modernist and romantic literature. At least since Wordsworth the sublime has always "brought the high and the low into dangerous proximity"; it "will always be found in the ill-defined zones of anxiety between discrete orders of meaning" (Weiskel 19, 21). In order for the hybrid narrative to *be* sublime, however, the "discrete orders" must be kept clear and separate. As Brian McHale points out in *Constructing Postmodernism*, the "high culture/low culture distinction" in particular, like the "Difference Engine" that provides the title of Gibson's and Sterling's one collaborative work, may well be necessary, "in however problematic or attenuated a form," in order for "the cultural engine to continue to turn over" (227).

McHale finds the cultural cross-fertilization in cyberpunk to be a characteristic as much of modernism as of postmodernism, but where the two may diverge is in the "technologically-enhanced speed of the traffic in models between the high and low strata of culture" (227). The postmodernist desire has been to erase the categorical distinctions that earlier forms of the sublime have traditionally depended upon. And nowhere is the loss of traditional distinctions more apparent than in cyberpunk's flattened conception of character and identity. The ultimate a posteriori criterion, once the last generation's highbrow productions are made operational and got ready for mass production, is functionality; and a new level of pragmatic functioning can be observed in cyberpunk's technological reconstitution of identity, what Gibson would call a "construct" or coded replication of a person's operational being. In *Neuromancer*, for example, characters who have died frequently persist—if their skills are deemed by those in power important enough to preserve—as pure disembodied information that, in a Turing test, would answer to all we ever knew of the living person. (Alan Turing proposed as a thought experiment putting questions to a computer; if the answers given could be human, the machine could be considered intelligent.) A person's expertise, valued solely for its ability to negotiate the complexity of the corporate network, could itself be accessed solely through a piece of patented software on a computer disk. No other human characteristic is necessary, for consciousness itself (Julian Jaynes and his four novelistic contemporaries notwithstanding) is an autonomous function that does not need to be imagined as embodied in flesh and language at all.[2]

Indeed, in *Neuromancer*'s adolescent mysticism, immediate experience through the body is mostly "a meat thing" to be rejected in favor of the cerebral pleasures and the "consensual hallucination" of cyberspace, Gibson's figure for the entire global network of electronic communications (51). Anything less than participation in the network is experienced by the cyberpunk hero as a fall into materiality, into the dreary consumerist world. Yet this hero hardly manages to escape this culture by trivializing the body; his abstract, disembodied, and "unthinkably complex" consciousness is not romantic transcendence, but is rooted in the very culture of high-tech consumerism that enables Gibson to rewrite the body on the abstract space of corporate capital.

2. In one sense, however, Gibson's cyberspace and the practical developments it has helped to inspire in virtual reality engineering do resemble Jaynes's theory. The cyberpunk's experience of concretely abstract spaces allows authors to present human identity as a kind of narrative in which the self becomes, in Jaynes's words, an " 'I' that can observe . . . space, and move metaphorically in it. [Consciousness] operates on any reactivity, excerpts relevant aspects, narratizes and conciliates them together in a metaphorical space where . . . meanings can be manipulated like things in space" (65–66). Compare James A. Connor's description of a future cyberspace where people "will be able to turn abstract objects like actuarial tables into physically manipulable virtual objects—shuffle them like cards or stack them like blocks. In cyberspace, information becomes the thing and the thing information" (7).

Its bodiless "data" are ultimately "made flesh in the mazes of the black market" (16), and its "money," a nearly autonomous signifier that has separated itself from any indexical reference to material value, is used precisely to generate "a seamless universe of self" (173).[3]

From the moment of its appearance, cyberpunk could be all too easily recognized as an aesthetic suited to the excesses and economic hubris of the Reagan era; and during a decade when a much-hyped research program in "artificial intelligence" was at last delivering not a simulation of embodied consciousness but numerous expert systems of corporate organization and control, cyberpunk's presentation of character could just as easily be attacked as "unnatural." Nonetheless, like traditional liberal humanist criticisms of postmodern writers for not creating "round" characters, objections to the informatics of cyberpunk identity may be ultimately beside the point. Such characterizations might be best understood, I think, not only as the dystopian expression of a cultural narcissism or of the oncoming hegemony of the machine, but as a logical response to the crisis of representation I have mentioned. The prospect that identity might become wholly informational enables Gibson, like Mailer, Pynchon, and their theorist contemporaries Fredric Jameson and Donna Haraway, to de-realize any notion of an individual and separate subject and thus to make identity itself an abstract representation of the vast and impersonal corporate networks that constitute so much of the contemporary lifeworld. As John Christie points out, Gibson, like Haraway, seems to have been "seduced by the possibilities of extending the technological sublime as the trope to figure out the political-aesthetic dimensions and possibilities of the infotech material formation" (180).

Thus, cyberpunk answered a specific strong challenge in the main postmodern debates of the mid-eighties. For even as *Neuromancer* was about to come on the scene, Jameson was closing one section of his benchmark essay "Postmodernism" with a partly depressed, partly exuberant suspicion that just such a "post-modern or technological sublime" was perceptible on the horizon (79). Writing in the *New Left Review* in 1984, Jameson could find in contemporary fiction (and not

3. Gibson takes monetary abstraction a step further in his second novel, *Count Zero*, where paper money—itself a base material symbol—is reserved almost exclusively for quasi-legal transactions in the criminal underworld. Ordinary transactions are accomplished solely by credit, which serves as passport and proof of citizenship no less than currency. As Kathy Acker writes in *Empire of the Senseless*, "money's flimsy paper people who don't have power carry on them" (32).

just science fiction) only "degraded" attempts to present the unpresentable, "faulty representations of some immense communicational and computer network," and "conspiracy theories" that at best revealed our inability to give concrete form to the abstract system of multinational capital (79, 80). That Jameson in this essay all but ignored the accomplishment of Pynchon, Coover, and Gaddis no less than McElroy, and that in an earlier essay he misapprehended a fledgling instance of the technological sublime in *Why Are We in Vietnam?*, has been less important for cultural theorists than his identification of a certain "dominant" in the terms of postmodern representation and narrative production. Given Jameson's subsequent inability, however, to offer much in the way of aesthetic or political alternatives to the dominant culture, we might look to these writers for possibilities that contemporary theory may have missed.[4]

In the very few pages of "Postmodernism" devoted to narrative fiction, Jameson was quite accurately describing not Coover's innovative "mainstream," which attacked the material structures, "priestly" truths, and "homologous form" of the society it sought to represent, but rather the majority fiction being published by an industry that, as the result of an unprecedented series of corporate mergers, was in the process of becoming an oligopoly.[5] Most fiction produced under such publishing conditions was bound to demonstrate all the vitiating postmodern symptoms listed by Jameson. These include a loss of critical distance from what the Frankfurt school used to call "the culture industry"; an assumption of the existence of a single hegemonic "system" rather than many possible and conflicting systems (such as impinge on the first-person consciousness of McElroy's protagonist in *Lookout Cartridge*); a flattening of sensory response to indistinguishable sources of cultural input (which Jameson aptly terms a "waning of affect" ["Postmodernism" 61]); and an impossibility of representing the technologically produced world system of multinational capitalism

4. Jameson's call at the end of "Postmodernism" for a "cognitive mapping" describes the problem of postmodernist representation, not its solution: in the absence of any single discourse system that would dominate all other systems, a one-to-one correspondence among systems might appear as undesirable as it is impossible.

5. Charles Newman is perhaps the only contemporary critic to look at the *aesthetic* role played by the "productive and social relations of the world which most affects" the writer, the world of corporate publishing itself (167). He cites the revealing statistic that "as of 1982, more than 50% of all mass market sales were accounted for by five publishers, and ten publishing firms accounted for more than 85%"—a situation that easily meets the accepted definition of an oligopoly (152).

from a position outside the system. Yet, Jameson's exclusive focus on institutional fiction (his one example of a postmodern "fabulist" being E. L. Doctorow), and his failure to note the innovative writing that major corporations happened to publish during the seventies and early eighties, may well have prevented him from seeing alternatives that were emerging from outside the publishing mainstream.[6]

Jameson, however, all but predicted cyberpunk, which for many observers has come to represent the ultimate logic of late capitalism, the very "apotheosis of postmodernism," in Istvan Csicsery-Ronay's quotable and often quoted phrase ("Cyberpunk" 182). Here was an aesthetic that could finally accept, and stylistically celebrate, its ineluctable position *within* the reproductive system of multinational capital. Its networks larger than life and circuits smaller than sight were no longer presented as *outside* the human mind and body at all. As in *Mao II*, where proliferating mass-reproduced imagery closes in on consciousness "like some computerized wall of enhanced sensation" (238), and as in *White Noise*, where the self is relentlessly put forward as nothing but "the sum of [its] data" (141), reality in a cyberpunk novel is always already a representation, a rich source of found language and ready-made fictions just waiting for the novelist to come along who is hip enough to pick up on them. (Reading this fiction, a friend once told me, is like going fossil hunting in the granite of a new McDonald's.) Indeed, it often seems as if cyberpunk's characters *cannot help* but represent to themselves the surrounding structure of mediations, simulacra, and machinic repetitions that have produced, for example, Baudrillard's simulation culture, Lyotard's postmodern sublime, or the dream space of Jameson's political unconscious. In John Christie's concise summation, the omnipresence of information has broken down

6. It is with some justice (and a good measure of self-service) that Curtis White takes Jameson to task for a 1992 collection that reprints the essay on postmodernism: "Why doesn't [Jameson] know," White asks, about the scene that had been emerging on the margins of corporate publishing in the preceding decade, "about Sun and Moon [White's publisher], City Lights's new fiction series, Dalkey Archive, Burning Deck, the Fiction Collective [which White helps direct], *Fiction International*, *Black Ice*, *Factsheet Five*, Semiotext(e), and *The Exquisite Corpse? Storming the Reality Studio?* . . . Let me ask you, Fredric Jameson, have you read Alan Singer's *The Charnel Imp?* Have you read Harold Jaffe's *Madonna and Other Spectacles?* Richard Powers's *Prisoner's Dilemma?* Gilbert Sorrentino's work with Dalkey Archive Press? Chandler Brossard? Gerald Vizenor's *The Heirs of Columbus?* . . . David Wojnarowicz? Marianne Hauser? Jerry Bumpus? Clarence Major? Rosaire Appel? Cris Mazza? George Chambers? Harry Mathews? Michael Ondaatje? Ursule Molinaro? Kathy Acker? Mark Leyner? None of this counts?" (21, 30).

"the modern's epistemological barrier between representation and the real" (180).

But the collapse of this barrier, like the purported collapse of the distinction between high and low culture, brings on problems of its own. For the challenge is not to collapse technological and aesthetic modes of production in yet another corporate merger but to integrate two distinct activities, to situate human consciousness, and to find, point by point in the nontotalizable network, those places where language converges with the real. A postmodern realism, such as Jameson among many others has sought but failed to recognize in contemporary fiction, will of necessity *be* sublime, though it need not reduce the unpresentable reality of the corporate system to a single image or web of paranoid connections. Unlike paranoia in its postmodern literary manifestations, a contemporary realism would not erase distinctions between mind and world, but would allow language to converge with systems of information and communication that must ultimately remain outside the representing mind and separate from the representation. To return again to my opening quotation of Henry Adams, the "universe that formed" the novelist (now perceived for better or worse as the system of multinational capitalism) must take shape in the novelist's "mind as a reflection of his own unity, containing all forces except himself" (475).

Unlike Adams, however, the cyberpunk hero is doomed to alternate—like Franz Pökler's alternation with servo-mechanical regularity between "personal identity and impersonal salvation" (*Gravity's Rainbow* 406)—between alienation and loss of self within the single, unthinkably complex system of world information. The cyberpunk's narrative remains the story not of integration but of an aggressively fragmented individual at war against a tabooed late capitalist totality. Indeed, the pervasiveness and ineluctable familiarity of its space of representation may explain why cyberpunk mostly sustains a generic narrative of romantic individualism, the most popular form of resistance from an earlier industrial moment (Jameson's and Ernst Mandel's "second stage" of modern industrial capital) which even apologists of cyberpunk would as soon have done with. As is often noted, to the embarrassment of those who would claim cyberpunk for some emergent literary avant-garde, when this fiction is not invoking traditional family structures against systemic technological domination, it frequently follows the popular pattern of the American detective hero—in Gibson's case a cyberspace cowboy—who must get his own back from a hostile class structure and diabolical political machine.[7] Admittedly, Pynchon and McElroy depended no less on this particular popular form: Oedipa Maas's quest for the Trystero in *The Crying of Lot 49*, like the search for Wintermute in *Neuromancer*, is essentially an epistemological quest that "reduces the intractability of reality to a figure, a kind of devilish entity that will stand for the spatial and informational chaos" that is the postmodern bureaucratic city (Siemion, "Whale Songs").

McElroy's questing detectives are rather less prone to suspect hidden plots and conspiracies; nor do his narratives ever lessen the complexity of technological systems, as Gibson and even Pynchon often do, by reducing them to a single paranoid figure. His heroes feel neither nostalgia for absent or receding centers (as in Pynchon) nor regret at their inability to bring events to closure. All of Cartwright's revelations, for example, are right there on the surface, and *Lookout Cartridge*'s maddeningly indistinct surfaces may be the nearest precursor to cyberpunk's determined and deliberate superficiality. Trenchcoat aside, Cartwright is less a hardboiled detective than a piece of software, "a computer virus," Siemion writes, who must gradually build himself back into a multinational corporate structure that he alters and contaminates without ever fully understanding ("Chasing the Cartridge" 135).

So, too, must Case, Gibson's hero in *Neuromancer*, learn to use a Kuang device—a slow computer virus capable of entering the technological structure and expanding the structure by virtue of its very presence—to penetrate a vast electronic library. Through the voice of the Flatline, a simulated construction of a once-living personality that exists precisely nowhere, Case is instructed in how to go beyond the crystalline logic of the mechanical structure: "This ain't bore and inject," says the Flatline. "It's more like we interface with the ice so slow, the ice doesn't feel it. The face of the Kuang logics kinda sleazes up to

7. Istvan Csicsery-Ronay provides the complete template for cyberpunk's "formulaic tales . . . in which a self-destructive but sensitive young protagonist with an (implant/prosthesis/technical talent) that makes the evil (megacorporations/police states/criminal underworlds) pursue him through (wasted urban landscapes/elite luxury enclaves/eccentric space stations) full of grotesque (haircuts/clothes/self-mutilations/rock music/sexual hobbies/designer drugs/teletechtronic gadgets/nasty new weapons/exteriorized hallucinations) representing the (mores/fashions) of modern civilization in terminal decline, ultimately hooks up with rebellious and tough-talking (youth/artificial intelligence/rock cults) who offer the alternative, not of (community/socialism/traditional values/transcendental vision), but of supreme, life-affirming *hipness*, going with the flow which now flows in the machine, against the spectre of world-subverting (artificial intelligence/multinational corporate web/evil genius)" (184).

the target and mutates, so it gets to be exactly like the ice fabric. Then we lock on and the main programs cut in, start talking circles 'round the logics in the ice. We go Siamese twin on 'em before they even get restless" (169).

What is perhaps most immediately striking about such a passage is Gibson's precise intuition of just how much abstract technology a modern audience can be expected to know. Nothing here would trouble even the school-age reader who is the traditional, though by no means exclusive, audience for science fiction—not the computer jargon, not the valorization of nonlinear and apparently nonrational processes *within* the province of a wholly technological operation, not even the concretization of abstract information in a crystalline lattice, expressed with minimal explanation by the homely figure of "ice" (from "*ICE*, intrusion countermeasures electronics" [28]). Gibson's least turn of phrase heralds an unprecedented familiarity not with technology itself but with its image. Case in cyberspace (along with Gibson and his readers in the era of home computers, graphic displays, user-friendly "icons," and system-breaking viruses like the Kuang) can take for granted a populist technology that earlier fictive heroes never had in such packaged and consumable forms. The merest mention of a "Sense/Net," an "Ono-Sendai," or a cyberpunk cowboy who "flatlined on his EEG" (50), to cite only random samples from the page before me—these things call up a ready-made aesthetic and reinforce a popular intuition of technological abstraction that earlier novelists had to *create* in their audience.

The relative ease of such reference may be not wholly a change for the worse. If, as I have suggested, the aesthetic potential of computer technologies has been materialized in surprisingly banal forms by the utilitarian technologies of our time, most of which are perhaps better suited to music and the visual arts than to the printed word, this materialization does not have to produce a degenerate form of narrative fiction. On the contrary, the new pervasiveness of technology may have created a more promising climate for the novel. Gibson's ready, even facile allusions demonstrate that technology may have become not simply another field of specialist discourse but something like the upscale, thoroughly bourgeois field of reference that has historically united an otherwise amorphous mass audience—whose existence has always been, since the mass production of books got fully under way in the nineteenth-century, both antagonist and mainstay of the realist and the naturalist novel.

The penetration of the corporate library itself in the passage just quoted may be read as an allegory of the compositional process that has preoccupied all of those postmodern writers who have sought to redefine realism in an age of expanding information. The technique spelled out by the Flatline is worthy of a classical rhetorician: one hides behind the mask of a powerful opponent, all the while diverting that opponent's language to other purposes. It is a way of poaching on power that has "momentum but not focus," in McElroy's formulation (*Lookout Cartridge* 79 passim). Also, in a way that again recalls Adams's position both within and outside the universe of force, the Kuang's operation dramatizes what can happen when a global electronic network of "unthinkable complexity" takes shape in the mind of a single protagonist—Case in *Neuromancer* (51), D.J. in Brooks Range, Brita in Beirut, Cartwright among the multinationals, or Imp Plus in the "matrix" of a reconstituted verbal and perceptual space. Gibson's strategy in this extraordinary passage is, in other words, possibly the most direct example yet of what I have been calling (with my own overlays onto Jameson, Lyotard, Leo Marx, and Weiskel) the technological sublime: the Kuang enacts a conceptual integration without a loss of personal identity, leaving the representing mind, as in Adams's sentence, separate from the complex of forces being represented.

But here the similarity between cyberpunk and its precursors ends, for what Gibson makes of this abstract space owes less to McElroy or DeLillo or the bourgeois community of nineteenth-century realism than to Pynchon at his most paranoid. As in *Gravity's Rainbow*, whose influence on *Neuromancer*, *Mona Lisa Overdrive*, and especially *Count Zero* Gibson gladly acknowledges, information in cyberworld comes to constitute the only real medium of exchange; and despite the Kuang strategy, power belongs not to any one person or group of people but to the computer networks themselves. Their coded mechanisms and visual patterns coming down through foreshortened generations confer "a kind of immortality" on the institutions "that have shaped the course of human history," while those who serve the institutions go on dying as they always have: "You couldn't kill a zaibatsu by assassinating a dozen key executives; there were others waiting to step up the ladder, assume the vacated position, access the vast banks of corporate memory" (Gibson 203).

This paranoid vision of a world totally connected and determined by corporate power could not be further from the multiple collaborative networks that McElroy imagines human society to be, in which the techno-scientist and the writer are engaged in independent thought adventures that rest on the same epistemological foundations. The corporate desire to "[transcend] the old barriers" (Gibson 203), which we have seen most clearly in the rocket scientist Wernher von Braun's compensatory vision of "the continuity of our spiritual existence after death" (*Gravity's Rainbow* 1), has its fictive counterpart in recurrent and varied post-transmortal projections: in Mailer's expiring consciousness at the start of *Fire* and throughout his fiction, in Pynchon's secular bureaucracies carried "beyond the zero" (*Gravity's Rainbow* 1), in McElroy's angelic hierarchies, and in DeLillo's Swiss hostage who has stumbled into the system of world terrorism. Ultimately, however, *all* such visions of technological transcendence, insofar as they promise to liberate us from our own materiality (in the earth, in our bodies, in the materiality of language), proliferate impersonal networks and corporate mergers that frustrate connections on a human scale. As in Pynchon's most dystopian enactments of the postromantic "Apollonian dream" (*Gravity's Rainbow* 754), the technology that permits such projections leads inevitably, apocalyptically, to what Norman O. Brown has called "deathly form" (157).

In an earlier chapter on *Gravity's Rainbow* I cited Kathy Acker's cheeky plagiarism of the passage just quoted from Gibson (a passage that is itself an indirect appropriation of Pynchon). Acker is certainly not the only and probably not the first certified postmodernist to engage with cyberpunk, but the directness of her engagement, and the peculiar twist she gives to its masculinist forms, has made her a member of the cyberpunk canon, a dot in the matrix. Says one character in *Empire of the Senseless*,

> "Interpersonal power in this world means corporate power. The multinationals along with their computers have changed and are changing reality. Viewed as organisms, they've attained immortality via biochips. Etc. Who needs slaves anymore? So killing someone, anyone, like Reagan or the top IBM executive board members, whoever they are, can't accomplish anything," I blabbed, and I wondered what would accomplish anything, and I wondered if there was only despair and nihilism. (83)

This passage from *Empire* is not the only borrowing from *Neuromancer*. Beginning with a section titled "Beyond The Extinction of Human Life" (31), Acker essentially rewrites Gibson's narrative, as in earlier novels she had rewritten (or at least resituated in a postmodern milieu) the narratives of Cervantes and of Dickens. (Acker even went so far as to title these earlier novels *Don Quixote* and *Great Expectations*, and to pen for their jackets admiring blurbs under the signature of "Alain Robbe-Grillet," no less.) This recycling of passages from classic and current writing is more complex than any particular lifting of a specific title, passage, or group of passages, and if "plagiarism" is perhaps a poor word for the practice, it is less misleading than a more apparently benign choice such as "influence." Despite the word's "negative connotations (especially in the bureaucratic class)," plagiarism may be an authentic aesthetic for the age. As the Critical Art Ensemble (nameless plagiarists all) argue, "This is the age of the recombinant: recombinant bodies, recombinant gender, recombinant texts, recombinant culture" ("Utopian Plagiarism" 85, 84). In a postmodern culture of proliferating categories and uncertain meanings, plagiarism becomes once more (as in the time before print technologies) a way of disseminating meaning across categories, of transmitting ideas to distant cultures or, as we would now say, discourse communities that would not otherwise get the word. Such an aesthetic might be sublime, except for a certain flatness, a tendency to present all texts as "reusable" and "all objects as equal, [horizontal] on the plane of phenomena" (88). A certain affectlessness comes to characterize the prose of the plagiarist, and without the possibility of heightened emotion, the work is, in the end, more strange than sublime.

Not that Acker, for one, would necessarily want to work the sublime register of her older male contemporaries. For Acker is not placing herself in the great tradition, nor is she seeking to establish herself as either the bad or the obedient daughter in the contemporary line of Pynchon and Gibson. Gibson's Kuang provides a fair analogy for Acker's overwriting and recombination of earlier narratives; but she would appear to be writing a more aggressively politicized version of Gibson's story and an "Elegy of the World of the Fathers" less nostalgic than Pynchon's, a narrative that would seek a female, if by no means conventionally "feminist," alternative to deathly form. Told in the voice of Abhor, a woman "who's . . . part robot, and part black" (3), Acker's cyborg narrative can appropriate earlier narratives because, as in Haraway, "the cyborg has no origin story in the Western sense"; and if Acker's heroes at times indulge traditional romantic fantasies of disembodiment and alienation, their narratives are perhaps not as likely as those of her precursors to demonstrate "the awful apocalyptic *telos* of the 'West's' escalating dominations of abstract individuation, an ultimate self untied at last from all dependency, a man in space" ("Manifesto" 150–51).

Acker is one of comparatively few women writers in America to have taken up Jameson's challenge to map ourselves cognitively onto the postmodern space produced by large systems of global power. But she refreshingly differs from her male peers and precursors in her refusal to take these systems so very seriously. Rather than worry about whether or not technology will come to dominate social and psychological life, she simply takes for granted—and takes *over* for her own various purposes—the great single narrative of disembodiment and masculinist domination. This Kuang-like assumption of the reality of corporate power, and the inclusion of its representations in her own mostly first-person stream-of-consciousness narrative, is no small form of resistance in itself. By simply rejecting a melodramatic presentation, Acker strips power of its *mystique*, in this respect resembling McElroy more than Adams, Pynchon, Mailer, or DeLillo.

Yet there may be a still more direct, and at first glance oddly traditionalist, resistance in *Empire*. Its aggressively obscene, visceral language would seem to be itself a way of returning readers to the reality —and repressed dangers—of direct bodily experience. Mailer used obscenity in much the same way in *Why Are We in Vietnam?* as an antidote to a consensus culture whose foremost representative, the corporation executive, "was perfectly capable of burning unseen women and children in the Vietnamese jungles, yet felt a large displeasure and fairly final disapproval at the generous use of obscenity in literature and in public" (*Armies* 60). Also as in Mailer, an aggressively sexual language is used (however ambiguously cross-gendered sexuality may be in Acker in comparison to Mailer) to recover a romantic narrative of wholeness, a conceptual unification and release into sensual freedom that can occur only, as in the following lyrical passage, when the soul is perceived to have gone out of the world. The speaker here is Thivai, Abhor's male "partner" and her co-narrator in *Empire:*

> Look, my sister: the eyes are gone. The Suns. No one's looking. You can now do whatever you want: Crying out; teasing the thickness of thighs; smouldering by smiling. Since the world has disappeared: there's nothing; no one looks at anyone.
>
> Since the world has disappeared: rather than objects, there exists that smouldering within time where and when subject meets object.

> This voluptuousness of your thighs. Odours seeping out of cunt juice and semen. Since the only mirrors are distorted; all is secret. Please come back to my arms. Without you I am nothing (38–39).

The objective world in this passage is resisted or rendered unreal by the imagining of physical love between two *subjects*, a process that is as deeply caught up in Thivai's lyrical desire for Abhor as it is for the "damsel" in Coleridge's "Kubla Khan" singing for her "demon lover" on Mount Abora.[8] Also as in Coleridge, the way to intersubjectivity in Acker depends as much on the materiality of language itself as on the physicality of sex, although Acker's particular grammar of love differs significantly from the romantic poet's. In a comparable passage in *Don Quixote*, David Brande reads love itself as "a transitive verb linking, but also separating subject and object" (199). As such, love sets up its own linguistic hierarchies, the sequence of subject-verb-object that needs to be deconstructed, no less than do traditional oppositions of mind-body or soul-body that the romantic writer tends to uphold. Thus, a language that "undercuts syntactic expectations, continually contracting subject and verb, for instance, or inserting awkward appositives" (Brande, 199), becomes in Acker a technology of nonorganic linguistic growth less like Mailer's in *Vietnam* than like McElroy's in *Plus*. (Brande himself relates Acker's narrative to Gilles Deleuze's and Felix Guattari's *Plus*-like conception of a "body without organs," a composite of "desiring machines" that interconnect with one another to form an immanent, never a transcendent, field of human subjectivity.)

Acker's own intersubjective "construct," pointedly named "Kathy" (34), is the "fence" that Abhor and Thivai have to reach in order to score the drug that will correct the "neurological and hormonal damage" inflicted on Thivai by unnamed corporate sponsors (31). A not-so-very hidden figure of the terrorist-author, "Kathy" belongs to an organization that seeks, in a possible echo of DeLillo, "the main system's end in white noise" (37). Acker's story, again, is straight out of *Neuromancer*, down to the incidental comment that "with enough endorphin analogue, Abhor [like Case's partner Molly] could walk on a pair of bloody stumps" (34; cf. Gibson 65).[9]

8. I do not mean to overstress high cultural reference in Acker's work. Sandra Bernhard's Lower East Side comedy *Without You I'm Nothing* was running when *Empire* appeared, and Bernhard is probably a presence in the passage I have cited.

9. In *Constructing Postmodernism* Brian McHale has helpfully cross-listed these and several other common passages in *Empire* and *Neuromancer*.

More significant than any shared content, however, is a common obsession with political terror, a topos that may have supplanted paranoia as the key structuring principle of postmodern narrative in the eighties and nineties. Like paranoia, literary terrorism may indicate nothing so much as the author's increased awareness of his or her own marginality. But where paranoia provided at least the appearance of control through the literary figuration of real plots centered on the single authorial consciousness, terrorism underscores the increasing difficulty of representing any reality that is not already simulated. "Imagination," writes Acker, "was both a dead business and the only business left to the dead. In such a world which was non-reality terrorism made a lot of sense" (35). Terror in Acker is finally, as in contemporary theorists such as Baudrillard and Lyotard, the price we pay for the persisting "transcendental illusion" that we can present reality as a totalized unity, least of all through any figure of a unitary author or autonomous, paranoically centered "text" (Lyotard, "Answering" 81).

Acker's recycling of earlier novels, then, can be understood as a form of resistance more direct even than DeLillo's recycling of his own and other postmodern texts in *Mao II*. In essence Acker has attempted to "cannibalize" or "techno-digest" (in a phrase Haraway cites from Zoe Sofia) "the organic, hierarchical dualisms ordering discourse in 'the West' since Aristotle" ("Manifesto" 163). Finally, however, this cannibalization would seem to follow the same dominant "logic" that Jameson identifies in "Postmodernism"; for what work demonstrates better than *Empire* the "new and heightened bricolage: metabooks which cannibalize other books, metatexts which collate bits of other texts," which Jameson identifies with "the general logic of postmodernism"(?)

In saying this I do not mean to indict cyberpunk fiction for failing to meet the standards of what I take to be its main literary antecedents. Such textual recycling is the mode not only of Acker and Gibson but of all those "most energetic postmodernist texts" that, according to Jameson, "tap the networks of reproductive process and thereby afford us some glimpse into a post-modern or technological sublime" (79). I would not be so quick as Jameson, however, to call such fictions "degraded," as if there were ever an unproblematical referent (some former "reality") that has somehow been lost in fiction since the early eighties. Even in Pynchon, the figure of paternalistic authority and source of so many popularly recycled images for the entire genre, we find a cannibalization of other, currently popular texts.

The example of Pynchon shows that any change in postmodern aesthetics is more than generational. As I have argued elsewhere (in "Pynchon's Groundward Art"), if *Gravity's Rainbow* engendered in cyberpunk writers a taste for technological complexities and multiple otherworldly bureaucracies, there are hints of cyberpunk all through Pynchon's later novel *Vineland*. The genre's reciprocated influence can be felt not only in details such as DL Chastain's ninjette training, but in a hybrid aesthetic, partly experimental, partly populist, that might best be understood as an attempt both to extend a modernist imperative for unceasing technical innovation and to retain some mooring in the shared social reality of nineteenth-century realism. This is how I have interpreted Pynchon's preference (in a statement that has irritated interpreters in the postmodern antirealist camp) for fiction that has some "grounding in human reality" and an authenticity "found and taken up, always at a cost, from deeper, more shared levels of the life we all really live" (*Slow Learner* 18, 21). If this shared life, in *Vineland* as in cyberpunk, is often hard to separate from the collective reality of brand names, corporations, and network television, Pynchon does imagine a full domestic life for his characters that was nearly absent from his earlier work. The new domesticity may be, however, like Gibson's disappointing reliance on Oedipal triangles and other conventional family narratives in the novels that followed *Neuromancer*, less a viable realistic ground for fiction than a reaction against the simulation culture that both novelists had anatomized in earlier work.

A desire to hark back to a reality less abstract than Baudrillard's simulacrum and less unthinkably complex than Jameson's "reproductive process" shows up in other small ways. There is a strangely dramatic moment near the end of *Neuromancer* when Gibson seems to retreat from "the unlimited subjective dimension" and invisible lines of cyberspace (63), when "whole stretches were being stripped back to steel and concrete" (204). Not coincidentally, this is precisely the passage lifted by Acker at the very moment when she interrupts her postmodern stream of consciousness (her "blab") to wonder whether, since it was not possible to "accomplish anything" by assassination, "there was only despair and nihilism":

> A boat was floating under the bridge. The body of an old man lay in the boat. I looked and saw this boat.
>
> "Old man," the black skeleton said. "He an important man. Whole stretches are being ripped to steel and concrete. Now he dead man."
>
> In the boat my father I had never known was dead. (83)

Thus, Acker joins Lyotard and Barthes in mourning the passing of monolithic modernist narrative and its authors. Whether or not some order other than the terrorist structure of "steel and concrete" might emerge to replace the phallocentric, Oedipal order of the dead Father, Acker is not saying.

A similar nostalgia for a literally "concrete" reality turns up in passages of *Vineland* that are easy to overlook. For example, in the 1940s the navy recruit Moody Chastain could find himself caught in the industrial spaces of a warship, "surrounded by nothing that did not refer, finally, to steel" (119). A decade later, when Moody's daughter (and Pynchon's contemporary) DL enters a fallout shelter that once housed "the Cold War dream," Pynchon can still have her imagine an "escape to some refuge deep in the earth, one hatchway after another, leading to smaller and smaller volumes" (255). The desire to occupy an ever narrowing cultural margin has always been strong in postmodern writing. But if *Vineland* tells us anything, it is that such an "underground" resistance has long since stopped being viable.[10] The repressive operations of earlier decades have been replaced in this novel by less visible arrangements instituted during the "Nixonian Reaction" and consolidated during the Reagan-Bush era, with the help of media images and technologies of surveillance that only *seem* to give clear and direct presentations of the world (239).

Neither Pynchon, Gibson, nor Acker can be faulted for the accuracy of their own multilayered and broadly intertextual presentations of the simulacrum culture. The apparently solid ground of domestic relations and concrete technology, however, is in itself no more real or tractable than any other text, as the author of *Gravity's Rainbow* knows very well. So when this author invokes the diminishing psychic and material spaces of "the Cold War dream," his nostalgia comes down again to a problem of referentiality, a desire for a narrative "weight" that an earlier fictive naturalism has traditionally achieved through a detailed romantic figuration of a concrete object world. Even such "crushing images as rolling mills and molten steel" at least provided the romantic imagination with a referential reality that could be fixed, and therefore resisted. But by now these products of the first industrial revolution have been brought under the control of " 'bits' in a flow of information traveling along circuits in the form of electronic impulses," as Italo Cal-

10. Steffen Hantke notes how, in *Vineland*, "political strategies grounded in the 1960s have become obsolete," and with them Pynchon's "strategy of invisibility" and the modernist "fiction of marginality as a means of empowerment" (177).

vino points out in *Six Memos for the Next Millennium:* "The iron machines still exist, but they obey the orders of weightless bits" (8).

The electronic simulacrum represents not a resistance to reality so much as its sublimation: abstraction enters the concrete material of the real, even as "lightness," a prime virtue in Calvino, appears as a "reaction to the weight of living" (26). In Calvino's aesthetic, although "lightness" is the preferred value, it is not opposed to "weight," any more than fantasies and dreams are opposed to the production of the real. Desire and the human imagination run through the weightiest machinery and the most disembodied electronic forms, and these things need the imagination no less than it needs them. The imagination gives technology the narrative form necessary for human significance, and technology, in whatever form, provides necessary referential constraints to the imagination.

Calvino, however, knows that he cannot turn to computer science to justify a literary viewpoint, any more than Pynchon or the cyberpunks can return to the steel and concrete of past technologies to make us aware of "the weight, the inertia, the opacity of the world" (4). Ultimately, fiction has its weight not in any particular figuration of technological content but in the materiality of language itself, the sensuous, referential movement of sentences and paragraphs that enables a novel, through all of its technical and linguistic innovations, also to *move* readers. Contemporary narrative does not necessarily move readers any less for invoking a reality that is ultimately unnamed and resistant to concrete reference. By all signs, the novel's affective power can only increase as the technological reality continues to exceed its figurations.

Works Cited

Abrams, M. H. "The Correspondent Breeze: A Romantic Metaphor." *Kenyon Review* 19 (1957): 113–30.
Acker, Kathy. *Empire of the Senseless*. New York: Grove, 1988.
Adams, Henry. *The Education of Henry Adams* (1918). Boston: Houghton Mifflin. 1961.
Adamowski, T. H. "Sex and Authenticity: D. H. Lawrence, Norman Mailer, and Saul Bellow." Unpublished talk given at the University of Toronto, January 1988.
Bakhtin, Mikhail. *The Dialogic Imagination: Four Essays*. Austin: University of Texas Press, 1981.
——— *Problems of Dostoyevsky's Poetics*. Minneapolis: University of Minnesota Press, 1984.
Barr, Marleen. *Feminist Fabulation: Space/Postmodern Fiction*. Iowa City: University of Iowa Press, 1993.
Barthes, Roland. "The Death of the Author" (1968). In *The Rustle of Language*, trans. Richard Howard. New York: Hill and Wang, 1986.
Bateson, Gregory. *Steps to an Ecology of Mind*. New York: Ballantine, 1972.
Baudrillard, Jean. *For a Critique of the Political Economy of the Sign*, trans. C. Levin. St. Louis: Telos, 1981.
——— *Selected Writings*, ed. Mark Poster. Stanford: Stanford University Press, 1988.
——— *Simulations*. In *Storming the Reality Studio: A Casebook of Cyberpunk and Postmodern Fiction*, ed. Larry McCaffery. Durham: Duke University Press, 1991.
Baym, Nina. "Melodramas of Beset Manhood: How Theories of American Fiction Exclude Women Authors." In *Feminist Criticism: Women, Literature, and Theory*, ed. Elaine Showalter. New York: Pantheon, 1985.
Benjamin, Walter. "The Work of Art in the Age of Mechanical Reproduction" (1936). In *Illuminations*, ed. Hannah Arendt. New York: Schocken, 1969.

Berman, Morris. *The Reenchantment of the World*. Ithaca: Cornell University Press, 1981.
Berressem, Hanjo. *Pynchon's Poetics: Interfacing Theory and Text*. Urbana: University of Illinois Press, 1993.
Bersani, Leo. "Pynchon, Paranoia, and Literature." In *The Culture of Redemption*. Cambridge: Harvard University Press, 1990.
Bertalanffy, Ludwig von. *General Systems Theory: Foundation, Development, Applications*. London: Alan Lane and Penguin Press, 1971.
Black, Joel D. "Pynchon's Eve of De-struction." *Pynchon Notes* 14 (1984): 23–38.
Blackmur, R. P. *Henry Adams*. New York: Harcourt Brace Jovanovich, 1980.
Bloom, Harold, ed. "Foreword" to Thomas Weiskel, *The Romantic Sublime*. Baltimore: Johns Hopkins University Press, 1986.
——— *Modern Critical Views: Samuel Beckett*. New York: Chelsea House, 1985.
Bradbury, Malcolm. "Closer to Chaos: American Fiction in the 1980s." *Times Literary Supplement*, May 22, 1992: 17–18.
Brande, David. "Making Yourself a Body without Organs: The Cartography of Pain in Kathy Acker's *Don Quixote*." *Genre* 24 (Summer 1991): 191–209.
Braun, Wernher von. "Why I Believe in Immortality." NASA press release, July 1969.
Brigham, Linda. "Motion and Destruction." Review of Paul Virilio, *The Aesthetics of Disappearance*. *American Book Review* 14 (June–July 1992): 10.
Brooke-Rose, Christine. "The New Science Fiction—Joseph McElroy: *Plus*." In *A Rhetoric of the Unreal: Studies in Narrative and Structure, Especially of the Fantastic*. Cambridge: Cambridge University Press, 1981.
Brooks, Peter. *The Melodramatic Imagination: Balzac, Henry James, Melodrama, and the Mode of Excess* (1976). New York: Columbia University Press, 1984.
Brown, Norman O. *Life against Death: The Psychoanalytical Meaning of History* (1959). London: Sphere, 1970.
Buck-Morss, Susan. *The Dialectics of Seeing: Walter Benjamin and the Arcades Project*. Cambridge, Mass.: MIT Press, 1989.
Bukatman, Scott. *Terminal Identity: The Virtual Subject in Postmodern Science Fiction*. Durham: Duke University Press, 1993.
Butler, Samuel. *Erewhon* (1872). New York: Penguin, 1987.
Cain, William E. "Making Meaningful Worlds: Self and History in *Libra*. *Michigan Quarterly Review* 29 (Spring 1990): 275–87.
Calvino, Italo. *Six Memos for the Next Millennium*. London: Jonathan Cape, 1992.
Carter, Dale. *The Final Frontier: The Rise and Fall of the American Rocket State*. London: Verso, 1988.
Cerf, Muriel. *Street Girl* (1975), trans. Dominic Di Bernardi. Normal, Ill.: Dalkey Archive Press, 1988.
Chomsky, Noam. *Manufacturing Consent: The Political Economy of the Mass Media*. New York: Pantheon, 1988.
Christie, John. "A Tragedy for Cyborgs." *Configurations* 1 (Winter 1993): 171–96.
Clerc, Charles, ed. *Approaches to "Gravity's Rainbow."* Columbus: Ohio State University Press, 1983.

Cohen, Sande. *Academia and the Luster of Capital.* Minneapolis: University of Minnesota Press, 1993.

Comnes, Gregory. Review of Susan Strehle, *Fiction in the Quantum Universe. Review of Contemporary Fiction* 12 (Summer 1992): 216–17.

Connor, James A. Review of Michael Benedict, ed., *Cyberspace: First Steps. American Book Review* 14 (June–July 1992): 7.

Coover, Robert. "The End of Books." *New York Times Book Review,* June 21, 1992: 1, 23–25.

——— "On Reading 300 American Novels." *New York Times Book Review,* March 19, 1984: 1, 37–38.

Critical Art Ensemble. "Utopian Plagiarism, Hypertextuality, and Electronic Cultural Production." In *The Electronic Disturbance.* New York: Automedia/Semiotext(e), 1994.

Csicsery-Ronay, Istvan. "Cyberpunk and Neuromanticism." In *Storming the Reality Studio,* ed. Larry McCaffery. Durham: Duke University Press, 1991.

——— "The SF of Theory: Baudrillard and Haraway." *Science-Fiction Studies* 18 (1991): 387–404.

DeCurtis, Anthony. Interview with Don DeLillo. *South Atlantic Quarterly* 89 (Spring 1990): 281–304.

Deleuze, Gilles. *Nietzsche and Philosophy* (1962). London: Athlone, 1983.

Deleuze, Gilles, and Felix Guattari. *Anti-Oedipus: Capitalism and Schizophrenia* (1972). Minneapolis: University of Minnesota Press, 1983.

DeLillo, Don. *Americana.* Boston: Houghton Mifflin. 1971.

——— "American Blood: A Journey through the Labyrinth of Dallas and JFK." *Rolling Stone,* December 8, 1983: 21–28, 74.

——— *Libra.* New York: Viking, 1988.

——— *Mao II.* New York: Viking, 1991.

——— *Ratner's Star.* New York: Knopf, 1976.

——— *White Noise.* New York: Viking, 1985.

Dornberger, Walter. *V-2.* New York: Viking, 1954.

Edmundson, Mark. "Romantic Self-Creations: Mailer and Gilmore in *The Executioner's Song. Contemporary Literature* 31 (Winter 1990): 434–47.

Federman, Raymond. "A Letter from the Galaxy" and "Cyberpunk Forum/Symposium," ed. Larry McCaffery. *Mississippi Review 47/48* 16 (1988): 35–39.

Fest, Joachim. *The Face of the Third Reich: Portraits of the Nazi Leadership,* trans. Michael Bullock. New York: Random House, 1970.

Fetterley, Judith. "'Hula, Hula,' Said the Witches." In *Critical Essays on Norman Mailer,* ed. J. Michael Lennon. Boston: G. K. Hall and Co., 1986.

Foreman, Joel. Review of Thomas Schaub, *American Fiction in the Cold War. American Quarterly* (March 1993): 176–86.

Foster, Hal. *Recodings: Art, Spectacle, Cultural Politics.* Port Townsend, Wash.: Bay Press, 1985.

Friedman, Alan J. "Science and Technology." In *Approaches to "Gravity's Rainbow,"* ed. Charles Clerc. Columbus: Ohio State University Press, 1983.

Gaddis, William. *JR.* New York: Knopf, 1975.

——— *The Recognitions* (1955). Cleveland: World, 1962.

Gibson, William. *Neuromancer*. New York: Ace, 1984.
Gleick, James. *Chaos: The Making of a New Science*. New York: Viking, 1987.
Goldstein, Laurence. *The Flying Machine and Modern Literature*. Bloomington: Indiana University Press, 1986.
Gontarski, S. E. "The Intent of Undoing in Samuel Beckett's Art." In *Modern Critical Views: Samuel Beckett*, ed. Harold Bloom. New York: Chelsea House, 1985.
Graff, Gerald. *Literature against Itself*. Chicago: University of Chicago Press, 1979.
——— *Professing Literature*. Evanston: Northwestern University Press, 1989.
Hantke, Steffen. *Conspiracy and Paranoia in Contemporary American Fiction: The Works of Don DeLillo and Joseph McElroy*. Frankfurt am Main: Peter Lang, 1994.
Haraway, Donna J. "The Actors Are Cyborg, Nature Is Coyote, and the Geography Is Elsewhere: Postscript to 'Cyborgs at Large.'" In *Technoculture*, ed. Constance Penley and Andrew Ross. Minneapolis: University of Minnesota Press, 1991.
——— "A Cyborg Manifesto: Science, Technology, and Socialist-Feminism in the Late Twentieth Century." In *Simians, Cyborgs, and Women: The Reinvention of Nature*. New York: Routledge, 1991.
——— " 'Gender' for a Marxist Dictionary: The Sexual Politics of a Word." In *Simians, Cyborgs, and Women: The Reinvention of Nature*. New York: Routledge, 1991.
Harris, Robert R. Interview with Don DeLillo. *New York Times Book Review*, October 10, 1982: 26.
Hayles, N. Katherine. *Chaos Bound*. Ithaca: Cornell University Press, 1990.
——— *The Cosmic Web: Scientific Field Models and Literary Strategies in the Twentieth Century*. Ithaca: Cornell University Press, 1984.
———, ed. *Chaos and Order: Complex Dynamics in Literature and Science*. Chicago: University of Chicago Press, 1991.
Hayles, N. Katherine, and Mary B. Eiser. "Coloring *Gravity's Rainbow*." *Pynchon Notes* 16 (Spring 1985): 3–24.
Heidegger, Martin. "The Question Concerning Technology" (1954). In *Basic Writings*. New York: Harper, 1977.
Heisenberg, Werner. *Physics and Philosophy: The Revolution in Modern Science* (1958). New York: Harper and Row, 1962.
Hite, Molly. "Feminist Theory and the Politics of *Vineland*." In *The Vineland Papers: Critical Takes on Pynchon's Novel*, ed. Geoffrey Green, Donald J. Greiner, and Larry McCaffery. Normal, Ill.: Dalkey Archive Press, 1994.
——— *Ideas of Order in the Novels of Thomas Pynchon*. Columbus: Ohio State University Press, 1983.
Hollinger, David A. Review of *Operation Epsilon: The Farm Hall Transcripts*. *New York Times Book Review*, December TK, 1993).
Ingendaay, Paul. *Die Romane von William Gaddis*. Trier: Wissenschaftlicher Verlag Trier, 1993.
Jameson, Fredric. "The Great American Hunter, or Ideological Content in the Novel." *College English* 34 (1972): 180–97.

——— *Marxism and Form*. Princeton: Princeton University Press, 1971.
——— *The Political Unconscious: Narrative as a Socially Symbolic Act*. Ithaca: Cornell University Press, 1981.
——— "Postmodernism, or the Cultural Logic of Late Capitalism." *New Left Review* 146 (1984): 53–92. Expanded and reprinted in *Postmodernism, or the Cultural Logic of Late Capitalism*. Durham: Duke University Press, 1991.
——— "Progress versus Utopia; or, Can We Imagine the Future?" *Science Fiction Studies* 14 (1982): 44–69.
Jaynes, Julian. *The Origin of Consciousness in the Breakdown of the Bicameral Mind* (1976). Boston: Houghton Mifflin, 1990.
Johnson, Christopher. *System and Writing in the Philosophy of Jacques Derrida*. Cambridge: Cambridge University Press, 1993.
Johnston, John. "'The Dimensionless Space Between': Narrative Immanence in Joseph McElroy's *Lookout Cartridge*." *Review of Contemporary Fiction* 10 (Spring 1990): 95–111.
Kant, Immanuel. "Analytic of the Sublime." In *Kant's Critique of Aesthetic Judgement*, trans. J. C. Meredith. Oxford: Oxford University Press, 1911.
Karl, Frederick. "American Fictions: The Mega-Novel." *Conjunctions* 7 (1985): 248–60.
Kármán, Theodore von. *The Wind and Beyond*. Boston: Little, Brown 1967.
Kittler, Friedrich. "Gramophone, Film, Typewriter," trans. Dorothea von Mücke and Phillippe L. Similon. *October* 41 (Summer 1987): 101–14.
Krol, Ed. *The Whole Internet: User's Guide and Catalogue*. Sebastopol, Calif.: O'Reilly and Associates, 1992.
Kucich, John. "Postmodern Politics: Don DeLillo and the Plight of the White Male Writer." *Michigan Quarterly Review* 27 (1988): 328–41.
Lacan, Jacques. "The Unconscious and Repetition." In *The Four Fundamental Concepts of Psycho-Analysis*. New York: Norton, 1981.
Laing, R. D. *The Divided Self*. New York: Random House, 1960.
Landow, George P. *Elegant Jeremiahs: The Sage from Carlyle to Mailer*. Ithaca: Cornell University Press, 1986.
——— *Hypertext*. Baltimore: Johns Hopkins University Press, 1992.
Landow, George, and Paul DeLany, eds. *Hypermedia and Literary Studies*. Cambridge, Mass.: MIT Press, 1991.
LeClair, Tom. *The Art of Excess: Mastery in Contemporary American Fiction*. Urbana: University of Illinois Press, 1989.
——— *In the Loop: Don DeLillo and the Systems Novel*. Urbana: University of Illinois Press, 1987.
——— "Missing Writers." *Horizon* (October 1981): 48–52.
LeClair, Tom, and Larry McCaffery, eds. *Anything Can Happen: Interviews with Contemporary American Novelists*. Urbana: University of Illinois Press, 1983.
Leed, Eric. "'Voice' and 'Print': Master Symbols in the History of Communication." In *The Myths of Information: Technology and Postindustrial Culture*, ed. Kathleen Woodward. Madison: University of Wisconsin Center for Twentieth-Century Studies, 1980.

Lennon, J. Michael, ed. *Critical Essays on Norman Mailer*. Boston: G. K. Hall and Co., 1986.

Lentricchia, Frank. "The American Writer as Bad Citizen—Introducing Don DeLillo." *South Atlantic Quarterly* 89 (Spring 1990): 239–44.

——— "*Libra* as Postmodern Critique." *South Atlantic Quarterly* 89 (Spring 1990): 431–53.

Leverenz, David. "On Trying to Read *Gravity's Rainbow*." In *Mindful Pleasures: Essays on Thomas Pynchon*, ed. George Levine and David Leverenz. Boston: Little, Brown, 1976.

Levine, George, and David Leverenz, eds. *Mindful Pleasures: Essays on Thomas Pynchon*. Boston: Little, Brown, 1976.

Limon, John. *The Place of Fiction in the Time of Science: A Disciplinary History of American Writing*. New York: Cambridge University Press, 1990.

Longinus. *On Great Writing (On the Sublime)*, trans. G. M. A. Grube. Indianapolis: Hackett, 1991.

Loose, Julian. Reviews of Don DeLillo, *Mao II*, and Frank Lentricchia, ed., *Introducing Don DeLillo*. *Times Literary Supplement*, August 30, 1991: 20–21.

Lyotard, Jean-François. "Answering the Question: What Is Postmodernism?," trans. Régis Durand. In *The Postmodern Condition: A Report on Knowledge*, trans. Geoff Bennington and Brian Massumi. Foreword by Fredrick Jameson. Minneapolis: University of Minnesota Press, 1984.

——— "The Sublime and the Avant-Garde" (1985). In *Postmodernism: A Reader*, ed. Thomas Docherty. New York: Columbia University Press, 1993.

Mailer, Norman. *Advertisements for Myself*. New York: Putnam, 1959.

——— "Alpha and Bravo." Unpublished manuscript, ca. 1969.

——— *An American Dream*. New York: Dial, 1965.

——— *Ancient Evenings*. Boston: Little, Brown, 1983.

——— *The Armies of the Night: History as a Novel, The Novel as History*. New York: New American Library, 1968.

——— *Cannibals and Christians*. New York: Dial, 1966.

——— *The Executioner's Song*. Boston: Little, Brown, 1979.

——— *Existential Errands*. Boston: Little, Brown, 1972.

——— *Harlot's Ghost*. New York: Random House, 1991.

——— *The Naked and the Dead*. New York: Rinehart, 1948.

——— *Of a Fire on the Moon*. Boston: Little, Brown, 1970.

——— *Pieces and Pontifications*, Boston: Little, Brown, 1982.

——— *The Prisoner of Sex*. Boston: Little, Brown, 1971.

——— *Why Are We in Vietnam?* New York: Putnam, 1967.

Maltby, Paul. *Dissident Postmodernists: Barthelme, Coover, Pynchon*. Philadelphia: University of Pennsylvania Press, 1991.

Mathews, Harry. "We for One: An Introduction to Joseph McElroy's *Women and Men*." *Review of Contemporary Fiction* 10 (Spring 1990): 199–226.

McCaffery, Larry. Interview with David Foster Wallace. *Review of Contemporary Fiction* 13 (Summer 1993): 127–50.

———, ed. "Cyberpunk Forum/Symposium." *Mississippi Review* 47/48 16 (1988): 16–45.

——— *Storming the Reality Studio: A Casebook of Cyberpunk and Postmodern Fiction*. Durham: Duke University Press, 1991.

McElroy, Joseph. "Fiction a Field of Growth: Science at Heart, Action at a Distance." *American Book Review* 14 (April–May 1992): 1, 30–31.

——— "Holding with *Apollo 17*." *New York Times Book Review*, January 28, 1973: 27–29.

——— *Lookout Cartridge* (1974). With a preface by the author, "One Reader to Another." New York: Carroll and Graf, 1985.

——— "Midcourse Corrections." *Review of Contemporary Fiction* 10 (Spring 1990): 9–55.

——— "Neural Neighborhoods and Other Concrete Abstracts." *TriQuarterly* 34 (Fall 1975): 201–17.

——— *Plus*. New York: Knopf, 1977.

——— "The Skylab Cluster." *Poly Prep Alumni Review* (Spring 1974): 2–6.

——— *A Smuggler's Bible* (1966). New York: Carroll and Graf, 1986.

——— Review of Henry S. F. Cooper, Jr., "13: The Flight That Failed." *New York Times Book Review*, March 11, 1973: 4–5.

——— *Women and Men*. New York: Knopf, 1987.

McGowan, John. *Postmodernism and Its Critics*. Ithaca: Cornell University Press, 1991.

McHale, Brian. *Constructing Postmodernism*. London: Routledge, 1992.

McHoul, Alec, and David Wills. *Writing Pynchon: Strategies in Fictional Analysis*. Urbana: University of Illinois Press, 1990.

McLuhan, Marshall. *The Gutenberg Galaxy* (1962). Toronto: University of Toronto Press, 1986.

Mendelson, Edward. "Gravity's Encyclopedia." In *Mindful Pleasures: Essays on Thomas Pynchon*, ed. George Levine and David Leverenz. Boston: Little, Brown, 1976.

Molesworth, Charles. "Don DeLillo's Starry Night." *South Atlantic Quarterly* 89 (Spring 1990): 381–94.

Moore, Thomas. *The Style of Connectedness: "Gravity's Rainbow" and Thomas Pynchon*. Columbia: University of Missouri Press, 1987.

Morrow, Bradford. Interview with Joseph McElroy. *Conjunctions* 10 (1987): 145–64.

Mosley, Nicholas. *Hopeful Monsters*. Elmwood Park, Ill.: Dalkey Archive Press, 1990.

Musil, Robert. *The Man without Qualities*. Vol. 1 (1930), trans. Eithne Wilkins and Ernst Kaiser. New York: Perigee, 1980.

Nadeau, Robert. *Readings from the New Book on Nature: Physics and Metaphysics in the Modern Novel*. Amherst: University of Massachusetts Press, 1981.

Nadotti, Maria. "An Interview with Don DeLillo." *Salmagundi* 100 (Fall 1993): 86–97.

Newman, Charles. *The Post-Modern Aura: The Act of Fiction in an Age of Inflation*. Evanston: Northwestern University Press, 1985.

Norris, Christopher. *Uncritical Theory: Postmodernism, Intellectuals, and the Gulf War*. Amherst: University of Massachusetts Press, 1992.

Pascal, Blaise. *Pensées*. London: Hammondsworth, 1966.

Penley, Constance, and Andrew Ross, eds. *Technoculture*. Minneapolis: University of Minnesota Press, 1991.

Poirier, Richard. *Mailer*. New York: Modern Masters, 1971.

——— "Venerable Complications: Literature, Technology, and People." In *The Renewal of Literature: Emersonian Reflections*. New York: Random House, 1987.

Porush, David. *The Soft Machine: Cybernetic Fiction*. New York: Methuen, 1985.

Pynchon, Thomas. *The Crying of Lot 49* (1966), New York: Harper & Row, 1990.

——— *Gravity's Rainbow*. New York: Viking, 1973.

——— "Is It OK to Be a Luddite?" *New York Times Book Review*, October 28, 1984.

——— "Nearer My Couch to Thee." *New York Times Book Review*, June 6, 1993: 3, 57.

——— *Slow Learner*. Boston: Little, Brown, 1984.

——— *V.* New York: Lippincott, 1963.

——— *Vineland*. Boston: Little, Brown, 1990.

Pyuen, Carolyn S. "The Transmarginal Leap: Meaning and Process in *Gravity's Rainbow*." *Mosaic* 15 (1982): 33–46.

Ramsey, Roger. "Current and Recurrent: The Vietnam Novel." *Modern Fiction Studies* 27 (1971): 415–31.

Redfield, Marc. "Pynchon's Postmodern Sublime." *PMLA* 104 (March 1989): 152–62.

Roth, Phillip. "Writing American Fiction" (1961). In *Reading Myself and Others*. New York: Farrar, Straus, 1975.

Schaub, Thomas. *American Fiction in the Cold War*. Madison: University of Wisconsin Press, 1991.

——— "Don DeLillo's Systems: 'What Is Now Natural.' " *Contemporary Literature* 30 (Spring 1989): 128–32.

——— *Pynchon: The Voice of Ambiguity*. Urbana: University of Illinois Press, 1981.

Schwanitz, Dietrich. "Systems Theory and the Environment of Theory." In *The Current in Criticism*, ed. Clayton Koelb and Virgil Lokke. West Lafayette, Ind.: Purdue University Press, 1987.

Seed, David. *The Fictional Labyrinths of Thomas Pynchon*. London: Macmillan, 1988.

Shklovsky, Viktor. "Art as Device." In *Theory of Prose*, trans. Benjamin Sher. Elmwood Park, Ill.: Dalkey Archive Press, 1990.

Siegel, Mark R. *Pynchon: Creative Paranoia in "Gravity's Rainbow."* Port Washington, N.Y.: Kennikat, 1978.

Siemion, Piotr. "Chasing the Cartridge: On Translating McElroy." *Review of Contemporary Fiction* 10 (Spring 1990): 133–39.

——— "Whale Songs: The American Mega-Novel and the Age of Bureaucratic Domination" (Ph.D. diss., Columbia University, 1994).

Stade, George. Review of McElroy, *Lookout Cartridge*. *Contemporary Literary Criticism* 5 (February 1975): 279.

Sterling, Bruce. "Preface" to *Mirrorshades*. In *Storming the Reality Studio: A*

Casebook of Cyberpunk and Postmodern Fiction, ed. Larry McCaffery. Durham: Duke University Press, 1991.
Strehle, Susan. *Fiction in the Quantum Universe*. Chapel Hill: University of North Carolina Press, 1992.
Sukenick, Ronald. Contribution to "Cyberpunk Forum/Symposium," ed. Larry McCaffery. *Mississippi Review 47/48* 16 (1988): 61–62.
Tabbi, Joseph. "The Compositional Self in William Gaddis's *JR*." *Modern Fiction Studies* 35 (Winter 1989): 655–71.
——— "Pynchon's 'Entropy.'" *Explicator* 43 (Fall 1984): 61–63.
——— "Pynchon's Groundward Art." In *The Vineland Papers*, ed. Donald Greiner and Geoffrey Green. Normal, Ill.: Dalkey Archive Press, 1994.
Tanner, Tony. *City of Words: American Fiction, 1950–1970*. London: Jonathan Cape, 1971.
——— "In the Lion's Den." In *Critical Essays on Norman Mailer*, ed. J. Michael Lennon. Boston: G. K. Hall and Co., 1986.
——— "Toward an Ultimate Topography: The Work of Joseph McElroy." *TriQuarterly* 36 (Spring 1976): 214–52.
Tillman, Lynne. "The Museum of Hyphenated Americans." *Art in America* 79.9 (September 1991): 55–61.
Todorov, Tzvetan. *Mikhail Bakhtin: The Dialogical Principle*, trans. Wlad Godzich. Minneapolis: University of Minnesota Press, 1984.
Tölölyan, Khachig. "War as Background." In *Approaches to "Gravity's Rainbow,"* ed. Charles Clerc. Columbus: Ohio State University Press, 1983.
Tolstoy, Leo. *Anna Karenina*. New York: Modern Library, 1950.
Vollmann, William T. *The Rainbow Stories*. New York: Atheneum, 1989.
Wallace, David Foster. "The Empty Plenum: David Markson's *Wittgenstein's Mistress*." *Review of Contemporary Fiction* 10 (Summer 1990): 217–39.
——— "E Unibus Pluram: Television and U.S. Fiction." *Review of Contemporary Fiction* 13 (Summer 1993): 151–94.
Weisenburger, Steven C. *A "Gravity's Rainbow" Companion: Sources and Contexts for Pynchon's Novel*. Athens: University of Georgia Press, 1988.
Weiskel, Thomas. *The Romantic Sublime* (1976). Foreword by Harold Bloom. Baltimore: Johns Hopkins University Press, 1986.
Weiss, Paul. *Principles of Development: A Text in Experimental Embryology*. New York: Holt, 1939.
Werenfels, Samuel. *A Dissertation Concerning Meteors of Stile, or False Sublimity* (1711); ed. Edward Tomarken. Los Angeles: William Andrews Clark Memorial Library, University of California, 1980.
White, Curtis. Review of Fredric Jameson, *Postmodernism, or The Cultural Logic of Late Capitalism*. *American Book Review* 14 (December 1992–January 1993): 21, 30.
Wiener, Norbert. *The Human Use of Human Beings: Cybernetics and Society* (1950). New York: Avon, 1967.
Wilson, Rob. *American Sublime: The Genealogy of a Poetic Genre*. Madison: University of Wisconsin Press, 1991.
Wilson, William S. "And/Or: One or the Other, or Both." In *Sequence (con)Sequence: (Sub)Versions of Photography in the 80s*, ed. Julia Ballerini. Annandale-on-Hudson, N.Y.: Aperture, 1989.

——— "Joseph McElroy and Field." Unpublished manuscript.
——— Untitled essay on Marjorie Welish's *Small Higher Valley* series of paintings, exhibited at the E. M. Donahue Gallery, New York, June 1993.
Wittgenstein, Ludwig. *Culture and Value*, ed. G. H. von Wright, in collaboration with Heikki Nyman. London: Blackwell, 1980.
Wolfe, Tom. "Stalking the Billion-Footed Beast." *Harper's Magazine* (November 1989): 45–56.
Wolfley, Lawrence C. "Repression's Rainbow: The Presence of Norman O. Brown in Pynchon's Big Novel." *PMLA* 92 (1978): 873–89.
Woodward, Kathleen, ed. *The Myths of Information: Technology and Postindustrial Culture*. Madison: University of Wisconsin Center for Twentieth-Century Studies, 1980.
Yeager, Patricia. "The 'Language of Blood': Toward a Maternal Sublime." *Genre* 35 (Spring 1992): 5–24.
Žižek, Slavoj. *The Sublime Object of Ideology*. London: Verso, 1989

Index

CPSIA information can be obtained
at www.ICGtesting.com
Printed in the USA
FSOW02n1108160217
30716FS

9 780801 483837